RA
566.27
.K33
2008

Kabat, Geoffrey C.

Hyping health risks.

WITHDRAWN

38489

$27.95

DATE

NW ARKANSAS COMMUNITY COLLEGE
LIBRARY
ONE COLLEGE DR – HWY 102
BENTONVILLE, AR 72712

BAKER & TAYLOR

HYPING HEALTH RISKS

HYPING HEALTH RISKS
Environmental Hazards in Daily Life and the Science of Epidemiology

GEOFFREY C. KABAT

COLUMBIA UNIVERSITY PRESS  NEW YORK

Columbia University Press
Publishers Since 1893
New York   Chichester, West Sussex

Copyright © 2008 Columbia University Press
All rights reserved

A Caravan book. For more information, visit www.caravanbooks.org.

Library of Congress Cataloging-in-Publication Data

Kabat, Geoffrey C.
Hyping health risks : environmental hazards in daily life and the science of epidemiology / Geoffrey C. Kabat
p. cm.
Includes bibliographical references and index.
ISBN 978-0-231-14148-2 (cloth : alk. paper)—ISBN 978-0-231-51196-4 (e-book)
1. Health risk assessment—Social aspects—United States
2. Epidemiology   3. Environmental health.   I. Title.

RA566.27.K33 2009
615.9′02—dc22

2008007285

♾
Columbia University Press books are printed on permanent and durable acid-free paper.
This book is printed on paper with recycled content.
Printed in the United States of America

c 10 9 8 7 6 5 4 3 2 1

References to Internet Web sites (URLs) were accurate at the time of writing. Neither the author nor Columbia University Press is responsible for URLs that may have expired or changed since the manuscript was prepared.

For Roberta and Daniel

and to the memory of Elvin A. Kabat

We, the scientific community, should take the blame for this. Toxicologists, and I'm one of them, have perpetuated the idea that if 100 molecules are going to kill you, then one molecule is going to kill 1 percent of somebody. We have a tremendous ability to analyze anything and everything.

—Michael A. Gallo

CONTENTS

PREFACE

We live in a society that conditions us to be hyperattuned to anything that may affect our health. In spite of dramatic improvements in longevity and medical care in recent decades, it seems that we feel our health and well-being are, if anything, more vulnerable to a myriad of external and internal threats. Doubtless, a major contributor to these feelings of vulnerability is the fact that we are constantly bombarded with news of novel health hazards that lurk in our everyday lives. Many of these hazards, even some that provoke regulatory action, have been greatly overblown. But somehow, the news announcing the demise, or the downgrading, of a hazard does not put us at ease to anywhere near the same extent that the announcement of a new hazard galvanizes our concern. How do certain low-level hazards get selected for attention, while others that may be of much greater consequence receive little attention? What are the factors—that is the professional specialties, social groups, institutions, and beliefs—contributing to the inflating of health risks? What are the consequences of this skewing of what is, after all, vital information for the public at large? Finally, on the most practical level, how are we to distinguish between exposures that have important effects on health and those likely to have minimal, or no, effect? These are some of the central questions this inquiry focuses on.

The idea for this book dates from the early 1990s, when I served on a committee convened by the U.S. Environmental Protection Agency on the question of the health effects of exposure to secondhand tobacco smoke. The committee's charge was to comment on a lengthy document that had been drawn up by the agency before it was approved for publication. As a member of the committee, I attended two meetings that were open to the public and at which all interested parties had an opportunity to give comments. In addition to the roughly fifteen panel members, there were numerous speakers representing the tobacco industry on the one hand and anti-smoking groups on the other, as well as interested individuals. What was most striking about the proceedings was how neatly the attendees were divided into two camps. Almost all of the speakers had a strong position either endorsing or criticizing the central conclusions of the report, i.e., that secondhand tobacco smoke was a "known human carcinogen" responsible for about 3,000 lung cancer deaths each year in nonsmokers. The members of the panel, to be sure, raised questions about specific aspects of the report. However, virtually no one, aside from the tobacco industry representatives, devoted attention to the weaknesses of the epidemiological studies of passive smoking on which its conclusions were largely based. Furthermore, no one questioned whether the report's conclusion, as stated, was scientifically justified. It was as if in this forum such questions were off-limits.

I had been invited to serve on the committee because, starting in the early 1980s, I had carried out studies of the health effects of both active and passive smoking as an epidemiologist working with Ernst L. Wynder at the American Health Foundation in New York City. As a medical student, in 1950, Wynder had published one of the first studies demonstrating that cigarette smoking was a cause of lung cancer, and he went on to become a major figure in preventive medicine and a lifelong crusader against smoking. As someone who had conducted one of the early studies on passive smoking and was familiar with the scientific literature on this topic, my position was that it was entirely plausible that exposure to secondhand tobacco smoke, or environmental tobacco smoke (ETS), could cause some additional cases of lung cancer and other diseases among people who had never smoked—especially those with heavy and prolonged exposure and with a constitutional susceptibility. Furthermore, I saw no rational justification for anyone to be exposed to this totally unnecessary and avoidable form of air pollution.

However, I was also aware that the average levels of secondhand smoke to which people were exposed were quite low compared to the concentrations that active smokers inhale into their lungs and that because of the low concentration—and the difficulty of knowing what an individual's exposure had been over a period of decades—the effects of exposure to ETS were likely to be at the limits of what epidemiologic studies could reliably detect. Put differently, we know that active smoking is associated with a greatly increased risk of lung cancer and other diseases, in part because smoking is a habitual behavior. Owing to its habitual nature, smokers can tell you, with some degree of accuracy, how many cigarettes they have smoked per day, on average, and when they started smoking, as well as if, and pretty much exactly when, they quit smoking. When it came to ETS, the exposure was much less concentrated and much more variable over time, and there was no way to measure an individual's exposure over a period of decades.

As a matter of scientific honesty, I felt that an official report by a government agency should include a discussion of the limitations of the available evidence and should provide all relevant information, even if this required that the conclusions be tempered. To give just one example, it was reasonable to expect a 500-page report on this topic to mention the fact that there was little indication that the death rate from lung cancer among women who had never smoked had increased over the preceding decades, as one might expect to see if passive smoking were a cause of lung cancer in nonsmokers. In my oral comments, I offered a number of similar points that were missing from the report. Although several prominent scientists at the hearings endorsed my comments, and in spite of the considerable resources expended on revising the document, none of these points was mentioned in the published version.

As a result of participating in the proceedings and of seeing these omissions, I could only conclude that, for all its many pages and tables, the report was essentially a political document designed to achieve a specific objective, namely to declare that secondhand tobacco smoke was an unambiguous hazard to be placed in the same category as such cancer-causing agents as tobacco smoke (when inhaled by an active smoker), ionizing radiation, asbestos, benzene, arsenic, and radon. Such a document would provide a powerful basis for local jurisdictions to pass ordinances limiting smoking in public places—something I wholeheartedly supported. My concern was that the failure to cite all of the relevant evidence and to temper its conclusion would damage the credibility of the EPA, hampering its effectiveness in the future. In addition, I felt that it would give the public a distorted impression of this particular hazard in relation to other hazards. (The willful interpretation of a position such as mine as a pro-tobacco industry stance is just one symptom of the dismal state of much of the discourse concerning health risks that is the focus of this book. It creates a climate in which crucial distinctions and shadings are not made and in which researchers and others dare not voice views that are attacked because they are not politically correct.)

I had assumed that the EPA's report on secondhand smoke was an isolated case. But soon after the committee's work came to an end, I read a letter to the *New York Times* by a professor at Yale University regarding the agency's draft report on the health effects of electromagnetic fields and was struck by how uncannily similar his criticisms of that report were to what I had observed on the topic of passive smoking. Slowly, it began to dawn on me that maybe there was a more general dynamic at work affecting how hazards get portrayed.

By the mid-1990s, widespread concern about possible environmental causes of breast cancer had led to the initiation of several large studies addressing this issue, and I became involved in two federally funded studies of breast cancer on Long Island. These were among the first epidemiologic studies to include a role for representatives of the community (breast cancer advocates) whose political activism had contributed to obtaining funding for the studies. The intense publicity these studies received—both locally and nationally—threw into relief many contentious issues regarding both the science pertaining to the study of a link between environmental pollution and breast cancer and the way in which the science gets interpreted for the public. First, scientific opinion was divided on the question of a likely role of environmental pollution in causing breast cancer, as well as on what specific pollutants to focus on and how to study them. Second, statistics indicating that breast cancer rates on Long Island in the late 1980s and early 1990s were slightly higher than the average for New York State—which was largely explainable on the basis of known risk factors—got transmuted into the "fact" that there was an "epidemic of breast cancer" on Long Island. In sum, the Long Island studies provided a fascinating window onto the way in which high-profile studies of health hazards were viewed by different parties and how a potential hazard could be inflated.

These experiences brought home to me just how susceptible information on health hazards is to being skewed and misinterpreted. The many parties involved—scientists, editors of medical journals, health and regulatory agencies, advocates, industry, the legal profession, and the media—are all, on occasion, capable of making their contribution to the distortion of health risks. Given the divergence in aims and interests of the various parties and the intensity of the stakes, few participants have stepped back in order to attempt to understand how information about what are important health risks can be so badly distorted. It always struck me that, as scientists, we had an obligation to allay unfounded fears and to clarify what is firmly known on a given question, what is reasonably suspected, and what is improbable. Putting various known and suspected risks in perspective should not conflict with undertaking studies of new potential factors, but all too often it does. This book is premised on the belief that the ways in which a number of prominent scares have been "manufactured" is worth examining.

***

I have benefited from the intellectual and moral support of colleagues and friends. Some provided extensive and detailed comments on the manuscript, took issue with particular judgments or facts, or pointed out important aspects that I had ignored. Others served as sounding boards for my efforts to make sense of a large body of diverse and difficult material. Beyond their specific comments, many conveyed a gratifying enthusiasm for what I was trying to achieve. I particularly want to thank Steven Stellman, David Parmacek, Robert Adair, James Hébert, James Enstrom, Gwen Collman, Dale Sandler, Roger Jenkins, John Neuberger, Gloria Ho, and Ruth Breslau. From our first contact, Patrick Fitzgerald, my editor at Columbia University Press, immediately understood, and believed in, the core idea for the book and encouraged me to take a broad and nuanced view of these controversies. My wife Roberta Kabat and our son Daniel have lived with this project since its inception. The importance of their unwavering support, demonstrated in ways both small and large, is hard to overstate.

ABBREVIATIONS

ACS	American Cancer Society
ALL	acute lymphocytic leukemia
BEIR	Biological Effects of Ionizing Radiation
Bq/m^3	Becquerels per cubic meter
BMJ	British Medical Journal
CHD	coronary heart disease
CI	confidence interval
CIAR	Center for Indoor Air Research
CPS I	Cancer Prevention Study I
CPS II	Cancer Prevention Study II
DDE	dichloro-diphenyl-dichloroethylene
DDT	dichloro-diphenyl-trichloroethane
ELF-EMF	extremely low frequency-electromagnetic fields
EMF	electromagnetic fields
EMFRAPID	Electric and Magnetic Fields Research and Public Information Dissemination
EPA	Environmental Protection Agency
ERT	estrogen replacement therapy
EPRI	Electric Power Research Institute
ETS	environmental tobacco smoke
GIS	Geographic Information System
HBV	hepatitis B virus
HPV	human papillomavirus
HRT	hormone replacement therapy
Hz	Hertz
IARC	International Agency for Research on Cancer
IEEE	Institute of Electrical and Electronics Engineers
IOM	Institute of Medicine
LIBCSP	Long Island Breast Cancer Study Project
mG	milligauss
μT	microtesla
NAS	National Academy of Sciences
NCI	National Cancer Institute
NIEHS	National Institute of Environmental Health Sciences
NIH	National Institutes of Health
NRC	National Research Council

OR	odds ratio
OSHA	Occupational Safety and Health Administration
PAH	polcyclic aromatic hydrocarbons
pCi/L	picocuries per liter of air
PCBs	polychlorinated biphenyls
Rn	radon
RR	relative risk
SEER	Statistics Epidemiology and End Results program
SERMs	selective estrogen receptor modulators
WHI	Women's Health Initiative
WHO	World Health Organization
WL	working level
WLM	working level month

HYPING HEALTH RISKS

1

INTRODUCTION
Toward a Sociology of Health Hazards in Daily Life

Each society shuts out perception of some dangers and highlights others. We moderns can do a lot of politicizing merely by our selection of dangers.

—Mary Douglas and Aaron Wildavsky, 1982

We are all familiar with what has been referred to as the "hazard *du jour*" phenomenon. Typically, it starts with media reports of the findings of a new scientific study indicating that some lifestyle behavior, consumer product, or environmental factor is linked to some dire disease. Coffee drinking is linked to pancreatic cancer. Eating chocolate is claimed to dispose to benign breast disease in women. Environmental pollution, we are told, may cause breast cancer. Studies appear to show a connection between exposure to electromagnetic fields from power lines and electric appliances and a host of diseases, starting with childhood leukemia. Use of cellular telephones may lead to brain tumors. Exposure to secondhand tobacco smoke, or passive smoking, is linked first to lung cancer, then to heart disease, and most recently to breast cancer. Silicone breast implants are associated with connective tissue disorders. The list could be extended at great length. Following the initial report, a second report may appear soon after yielding further suggestive evidence of a hazard or, just as often, showing no effect. In this way, over time, a scientific literature develops on each topic marked by weak and inconsistent results, and the perception of a hazard takes on a reality.

Some scares, such as those surrounding coffee and cell phones, may subside fairly quickly, as better studies are published or as the hazard is put in perspective and deflated. In other cases, the hazard can take on a life of its own and persist over years or decades, becoming the focus of scientific research, regulatory action, lawsuits, and advocacy campaigns. In the case of electromagnetic fields from power lines, tens of billions of dollars have been spent to remediate a problem whose very existence is uncertain. But what characterizes all of these scares is that the public's perception of a hazard was greatly exaggerated and was not counterbalanced by an awareness of the tenuousness of the scientific evidence or of the relatively modest magnitude of the potential risk. Thus, to a large extent, when one examines the public's response to a high-profile health scare, one is dealing with the dissemination of poor information and appeals to fear. (It sometimes seems that the intensity of the fear is inversely proportional to the actual magnitude of the threat.)

To be sure, reporting of hazards in the media tends to reduce any question to the simplistic message that exposure X may cause condition Y, and the extensive background and necessary qualifications are rarely provided. But media attention to health hazards is only a symptom of a pervasive preoccupation in our society with risks to our health, and to understand their significance, one has to examine both the context and the full range of contributing factors. The distortion and inflation of health hazards is the result of a complex interplay between the consumers of information about health risks (the media, the public, activist groups) and the producers of this information (scientists and regulators), and the influences between these groups flow in both directions.

The processes and mechanisms by which the available scientific evidence on a question can be distorted have received little attention. This is not surprising. For one thing, studies in the areas of epidemiology and environmental toxicology are highly technical, and often there is disagreement among scientists about the available evidence and about the importance of a given agent, making it difficult for nonspecialists to evaluate. Furthermore, practicing scientists may not be inclined to look back and assess their own work critically or to examine the factors that led them to devote attention to a particular question as opposed to another. If one line of work does not pan out, scientists move on to something new, which is, by definition, "exciting" and "promising." Science, like other disciplines, has its own internal structure of incentives and rewards that keep practitioners focused on the goals of improving methods and conducting more rigorous and powerful studies of questions that are agreed at a given point in time to be important. There is little incentive, therefore, to question the value of research on a specific topic, especially when that topic is of great concern to the public and regulators. But there are other reasons for the failure to examine certain health risks objectively, which we will come to in a moment.

In addition to practitioners directly involved in public health, psychologists and others have devoted attention to the new academic disciplines of "risk perception" and "risk communication." Their objective has been to describe how people interpret information about various risks and how this calculus influences personal behavior and public policy. While contributing valuable insights, work in this area generally ignores how science is actually carried out, the impact of external factors on science, and the context in which its findings are interpreted and communicated to the public—precisely the issues that are central to this book. It is striking that some of the key collections of papers on risk perception make virtually no mention of epidemiology.[1] This can be explained by the fact that these new disciplines focus on the *consumers* of information regarding risk rather than on its *producers*.

The purpose of this book is to elucidate how confusion about what is a threat comes about and what factors—both internal and external to science—contribute to it. How do certain perceived health hazards get blown out of proportion, in spite of well-known facts that should help to keep irrational fear within bounds? In attempting to address these questions, I have set myself three distinct but interconnected aims. First, I try to sketch out the variety of factors and processes that contribute to the manufacture of a hazard. Second, by introducing the reader to the basics of the science of epidemiology, I demonstrate the difference between well-established, important hazards and small, uncertain, or hypothetical hazards. Finally, in four detailed cases studies I attempt to show how particular hazards have been inflated.

Health hazards of the kind that get seized on by the media, studied by different scientific disciplines, addressed by regulatory agencies, contested in the courts, and embraced by advocacy groups are what the nineteenth century father of sociology Émile Durkheim termed "social facts." That is, they are reflections of a particular society, and they function in specific ways within that society. To understand how certain health risks become greatly exaggerated, what is needed is an approach that incorporates an examination of the context in which science is carried out in the area of public health and how pressures and agendas that are internal as well as external to science can influence both what is studied and what is made of the scientific evidence. In other words, what is needed is a sociology of science in this area. This chapter attempts to describe the landscape in which certain health risks have been selected and distorted. Specifically, I examine the different groups and institutions that have played a role in the manufacture of several prominent hazards, as well as key considerations that have frequently been lost sight of in the public, and, even in the scientific, discussion.

THE RISE OF EPIDEMIOLOGY AND ENVIRONMENTALISM

It is a paradox of contemporary life that, although longevity as well as general health and well-being have increased dramatically in the United States over the past hundred years, we live in a society that is hyperattuned to any threat to our health, no matter how hypothetical or how small. This heightened concern about potential threats to our health stems from at least two major developments dating from the 1960s and 1970s: the rise of the field of epidemiology, the science of the determinants of disease in human populations, and the more or less concurrent rise of the environmentalist movement. Together, these two developments have not only created a new awareness of the determinants of health, but they also continuously reinforce and feed this consciousness with novel findings that command attention. Beginning with the first studies linking cigarette smoking to lung cancer and other diseases in the 1950s, findings from epidemiology have made commonplace the notion that specific behaviors and exposures may affect health. Owing to these discoveries, a new and powerful paradigm has emerged, offering the prospect that many common chronic diseases could be drastically reduced by curtailing the causative exposures. Over the same time period, the recognition of the need to regulate environmental pollution—embodied in the creation of the federal Environmental Protection Agency (EPA)—and the publicity surrounding many environmental issues—from air and water pollution to climate change—have fostered an unprecedented and widespread perception that our actions have effects on the environment and that the environment has effects on human health. While reports of specific findings may be difficult for the lay public to assess, each new report of an environmental hazard or of a putative link between an exposure and disease underscores the message that we are surrounded by myriad agents in the environment that may pose a threat to our health.[2]

Today the science of epidemiology occupies such a prominent place in the health sciences, in regulatory affairs, and in the news that it is easy to forget how recent a development this is. When the landmark studies linking cigarette smoking with lung cancer appeared in the early 1950s, there were no departments of epidemiology, and the methods used in these early studies were untested and were challenged by some of the most respected statisticians. During the 1960s and 1970s, epidemiology proved itself as a science by a series of important successes in identifying preventable causes of disease.[3] Over the past four decades the field has grown at a prodigious rate, whether measured by the number of departments and programs in schools of public health, the number of textbooks devoted to subfields of epidemiology, or the number of research papers appearing in journals.

Since epidemiology is a young science and because, of necessity, it relies for the most part on observational, as opposed to experimental, studies, it is not surprising that many of the associations that are found and reported turn out to be spurious, that is, noncausal, associations. For one thing, chronic diseases like cancer, heart disease, and Alzheimer's are complex, multistage, "multifactorial" conditions. (Lung cancer, which has one predominant cause—smoking—that accounts for the vast majority of cases, is probably an exception. For example, a large, international study has recently shown that ten different factors make a contribution to the occurrence of heart disease.) Furthermore, the relevant period of exposure for "diseases of long latency" like cancer and heart disease can be decades preceding the appearance of symptoms, making it difficult to establish a causal relationship. Usually, one has only crude information collected at one point in time to represent the actual exposure over this long period. Thus, a major problem confronting epidemiologic studies of low-level environmental exposures that have been the focus of public concern is that it is extremely difficult to know what an individual's exposure was over a period of decades that may be relevant to his or her risk of developing disease.

Finally, many of the strong relationships—the low-hanging fruit, so to speak—like smoking and lung cancer and alcohol consumption and oral cancer have already been identified. It is much more difficult to reliably identify new factors that play a role in a given disease when these may be subtle, leading at most to a doubling or tripling of the risk in those with the factor compared to those without the factor. Often, we are dealing with even smaller increases of 10–100 percent. (For comparison, people who have ever smoked have a tenfold, or 1000 percent, increased risk of lung cancer; current smokers have a 20-fold risk; and heavy current smokers can have as much as a 50- or 60-fold increased risk.) Scientists who have devoted their careers to studying the possible impact of low-level environmental exposures on chronic diseases readily acknowledge the immense difficulty of establishing credible linkages.[4] Thus, it has to be realized that many more potential "risk factors" are studied than turn out in the end to be causes of disease. This is inherent in the process of science and especially a young and nonexperimental science such as epidemiology. This basic truth is well known to any scientist, and yet when it is formulated explicitly and vigorously, it comes as a surprise.[5] It seems, therefore, to conflict with deep, unconscious beliefs.

For these reasons, whether a hazard exists and, if so, its magnitude is often difficult to resolve. In some instances the vague penumbra of a hazard associated with a given exposure can persist for decades—despite, or because of, inconclusive evidence—contributing to a pervasive sense of distrust regarding the environment. Over time, the succession of media reports raising the possibility of new potential hazards has the contradictory effects of reinforcing a general message—the importance of "lifestyle" behaviors and the environment on health—and at the same time thoroughly confusing the public about what to believe.[6] It is easy to see how, in many people, this torrent of "significant" findings would produce only a deep and generalized sense of anxiety or else apathy and fatalism. And such fatalism may itself be a health risk![7]

I should note that the vast majority of research on factors influencing health never attracts the attention of the public and usually is of interest only to a small number of specialists working in a particular area. It is only when research pertains to a potential hazard that has become a focus of intense public concern and government action that the kinds of distortion and misrepresentations that I will discuss take place.[8]

THE SCIENCE CONCERNING ENVIRONMENTAL HAZARDS IS CONTESTED

Although we tend to think of science as occupying an Olympian realm that is insulated from social, political, and ideological influences, the fact is that science is not conducted in a vacuum. Historians and sociologists of science like Thomas Kuhn, Paul Feyerabend, Robert K. Merton, and Michel Foucault have drawn attention to the role of historical, social, ideological, and personal factors in the formation and acceptance of scientific ideas.[9] Particularly in the area of science dealing with health and the environment, different groups and institutions have very different interests at stake when it comes to the science and policy concerning health risks. The starting point for any discussion of external influences on science in this area has to be industry's long-standing and well-documented record of efforts to suppress or neutralize credible scientific evidence of adverse health effects due to its products or processes. Tobacco, asbestos, and lead are only some of the more dramatic and better known entries in this record.[10] Powerful industries have routinely used their substantial resources in the legal, regulatory, and public relations spheres to put a favorable spin on the science relating to their products and to thwart regulation.[11] Furthermore, how vigorously regulatory agencies, like the Environmental Protection Agency and the Food and Drug Administration, confront powerful corporate interests varies with the political climate. We have recently witnessed a powerful demonstration of how the party in power can attempt to edit and mould scientific findings to fit its philosophy and the requirements of its political supporters.[12]

Given the inherent difficulty of establishing credible links between low-level exposure to environmental toxins and chronic disease, it is hardly surprising that the assessment of potential environmental health hazards is hotly contested and that there is a sharp antagonism between the proponents of unfettered freedom for commercial enterprise and those concerned with protecting the public's health and improving occupational safety. Each side tends to cite the evidence that supports its point of view in order to influence public policy.

It needs to be recognized that, in spite of industry's capacity to manipulate science affecting its interests, those who present themselves as defenders of public health and the environment are not always free of bias and self-serving agendas. But, whereas most of us are well aware of industry's bias, we are much less on guard against other manifestations of bias from other quarters, especially when these are cloaked in a self-validating rhetoric. The highly charged climate surrounding environmental health risks can create a powerful pressure for scientists to conform and to fall in line with a particular position. Researchers who fail to find evidence of an adverse health effect or who criticize the reigning consensus on a particular topic can be branded as biased in favor of industry.[13] When it comes to questions about the long-term health effects of low-level environmental exposures, the scientific evidence is often ambiguous, and policy decisions have to be made on the basis of incomplete data. In this situation, what is at stake far exceeds the often inadequate evidence on a specific question. Owing to the intensity of the conflicting points of view, a distorted picture of the evidence on a given question can be built up, and few in a position to know will risk the costs of speaking out. Furthermore, few attempts are made to step back and attempt to sort out what the scientific evidence has to say, as opposed to ideologically or professionally driven interpretations. For these reasons, things are not always what they appear to be when we are dealing with topics on which deeply entrenched interests clash.

If I have focused on examples of hazards that have been inflated, this is partly because this phenomenon has received little critical attention, in contrast to instances in which evidence of real or probable hazards has been suppressed. The fact that a threat was overblown, and how this came about, is simply not newsworthy in the way that a new alleged hazard is. What was the focus of great concern at one point in time merely fades from view and is replaced by new claims on our attention. However, I believe that these episodes have much to tell us about how scientific evidence can be overplayed and distorted information conveyed to the public. Furthermore, I believe that the tendency to overstate the evidence, for whatever purpose, actually strengthens the opposing party's hand. It sanctions the partisan use of science that should be rejected, no matter who is engaging in it. Ultimately, by failing to put certain potential hazards in perspective, one confuses the public and diverts attention from issues that may be far more important. Certainly, there is no dearth of environmental and public health problems that need sustained—not rhetorical—attention.[14]

PERCEPTION OF RISKS BY THE PUBLIC AND BY SCIENTISTS

There is a large gap between the way in which the general public perceives environmental risks and the way scientists view these same risks. For example, in a survey conducted by Harvard University, 55 percent of lay women believed that "chemicals" in the environment played a role in causing breast cancer, whereas only 5 percent of scientists did.[15] Other surveys have shown that "technical experts" and the public have different perceptions and "acceptance" of risks from diverse sources of radiation exposure.[16] For example, technical experts assessed the risk associated with nuclear power and nuclear waste as "moderate" and "acceptable," whereas the public viewed it as "extreme" and "unacceptable." Scientists also rated food irradiation as "low risk" and "acceptable," whereas the public rated food irradiation as "moderate to high risk." In contrast, the public was "apathetic" toward residential radon, which it considered "very low risk," whereas experts judged it to carry "moderate risk" and to require "action."

Consideration of benefits, as well as risks, also influences the perception of hazards. Results of a national survey regarding perception of risks and benefits from radiation and from chemicals indicated that risks from nuclear power were viewed as moderately high and the benefits low, whereas the risks from medical X-rays were viewed as low and the benefits moderately high. Similarly, risks from pesticides were seen as moderately high and the benefits low, whereas prescription drugs were viewed as posing a low risk and having large benefits.[17] Researchers have discerned a logic behind these subjective perceptions of risk. Certain risks, such as those associated with X-rays and prescription drugs, are accepted because of their perceived benefits and because of the trust placed in the medical and pharmaceutical professions. In contrast "the managers of nuclear power and non-medical chemical technologies are clearly less trusted and the benefits of these technologies are not highly appreciated, hence their risks are less acceptable."[18]

In addition to such subjective factors influencing the perception of risk, basic concepts can have one meaning for the specialist and a very different meaning in popular usage. In popular usage, the word "risk" immediately suggests a hazard, whereas for the scientist it denotes a purely mathematical statement about the occurrence of a phenomenon. Similarly, the word "environment" as used by scientists means something very different from what it means in common usage.[19]

BELIEF AND ADVOCACY

Anyone confronted with an inexplicable and serious disease will have a powerful motivation to find an explanation. And, since the mind abhors a vacuum, a *possible* explanation is preferable to no explanation. When families and advocacy groups focus on a particular condition about which little is known, the drive to come up with an explanation can be so powerful that it can lead people to disregard solid scientific evidence and opinion in favor of a plausible-sounding explanation that has no scientific support. A recent instance of this phenomenon is the widespread conviction among many parents of children with autism that thimerosol, a mercury-containing preservative, used in some childhood vaccines must have played a role in their children's disease, despite the existence of large studies that show no support for this claim.[20] Similarly, some breast cancer activists have held a strong conviction that environmental pollution must play a major role in explaining breast cancer occurring in their communities, although what is known about breast cancer does not suggest that pollution plays an important role. Popular movies like *A Civil Action* and *Erin Brokovich* reflect the powerful drive to identify the cause of a cluster of cases of a rare disease, even though the science is often far from conclusive.[21] The sheer force of belief is also in evidence among antismoking activists who have made questionable claims about the adverse health effects of modern smokeless tobacco.[22]

Depending on the specific characteristics of the hazard (in terms of perceived risks and "acceptance") and the nature of the available scientific evidence, the existence of a hazard can be more or less widely—and uncritically—accepted by the general population and sometimes by scientists themselves.

MEASUREMENT, PARACELSUS, AND LOW-LEVEL EXPOSURES

Modern science can measure contaminants down to the parts per billion range and below; however, it is much more difficult to determine what levels of exposure are associated with long-term health effects.[23] Extremely low levels of dioxin can be detected in paper towels, and methylmercury has been detected in farm-raised salmon. Of course, we should not be complacent about the presence of detectable toxins in our food and environment, and it is imperative to rigorously monitor the levels of toxins in the environment, in food, and in our bodies. At the same time, it has to be recognized that health scares can be triggered by reports of what are extremely low levels of a toxin that are well below the level likely to have any biological significance.[24] In the case studies discussed in this book this basic distinction either tends to get lost or, often, to be totally absent. Furthermore, it also has to be realized that the cost of eliminating every trace of toxic compounds from food and the environment may make it unfeasible; and even if this could be achieved, such a step might entail other negative consequences, such as making food less affordable.

In public discussions of environmental hazards, the critical principle of toxicology, attributed to the sixteenth century alchemist and physician Paracelsus, that "the dose makes the poison" is rarely even acknowledged. And yet, it is well-established that there are many substances that are toxic at high levels but beneficial at low levels. This is true of essential nutrients, including vitamin A, beta-carotene, and trace elements such as iron, copper, zinc, selenium, cobalt, magnesium, and manganese. It is also true of ethyl alcohol. But whether there is a benefit, no effect, or a harmful effect, when we get down to very low levels, often the focus is only on the *existence of a risk*, without any accompanying awareness of how low the exposure is. Often we are operating in a low-dose region where it is difficult to determine what the effect is, or whether there is any effect. In some instances the lack of definitive resolution regarding the health effects of very low-level exposure to an agent such as ionizing radiation can paradoxically lead concerned groups to ignore just how low the levels actually are.[25]

Few people are aware of the extent to which certain controversies can be about exquisitely small doses and effects. It should be noted that scientists bent on promoting the importance of their work or on furthering an ideological cause can also fail to acknowledge this principle.

SCIENTISTS

Scientists can contribute to the manufacture of a hazard or its inflating in a number of ways and for a number of reasons. Being human, scientists can get excited about a novel result that could have important implications, and they may overstate the significance of their findings. The familiar pattern of a dramatic new finding that is not confirmed when more careful studies are carried out is no doubt partly explained by the desire for novel findings. Furthermore, scientists can develop an attachment to a particular hypothesis or line of inquiry, and this can subtly influence the way in which they evaluate the relevant evidence.[26] Sander Greenland, an epidemiologist at the University of California, Los Angeles, has provided an incisive formulation of what is at issue when interpreting research findings:

> There is nothing sinful about going out and getting evidence, like asking people how much do you drink and checking breast cancer records. There's nothing sinful about seeing if the evidence correlates. There's nothing sinful about checking for confounding variables. The sin comes in believing a causal hypothesis is true because your study came up with a positive result, or believing the opposite because your study was negative.[27]

This applies not only to the individual study but also to the totality of the evidence on a given question. In both cases it is imperative to critically evaluate all of the relevant evidence and to assess what can be concluded from it. But the fact remains that it is difficult to study a new question without in some way privileging those findings that tend to support the hypothesis. There is a built-in disposition to want to believe that what one is studying is important.

In order to obtain funding, a scientist needs to make a convincing case to other scientists—who are his peers and competitors—that his idea is worthy of receiving limited federal dollars. And to be successful, it is crucial for a grant proposal to contain "preliminary data" that lend support to the proposed line of research. This can create a pressure toward emphasizing what is "interesting" in one's recent results in order to make the strongest case for the importance of the proposed work. Science is no different from other more mundane fields of endeavor—one has to *sell* one's idea.

Scientists can also overstate the importance of a body of evidence in order to further a policy objective that they feel is important. Furthermore, once an inflated assessment becomes widely accepted, there is an incentive to emphasize confirmatory evidence and downplay other evidence.

The real-world conditions under which epidemiological research on environmental factors affecting human health is carried out also affect how questions regarding hazards are initially conveyed to the public and pursued. Attention to a new environmental hazard tends to conform to a pattern. An early, provocative study—which is usually small and rudimentary—appears to show evidence that a given exposure is associated with a given disease. In response to this initial finding, scientists undertake new studies to address the question. In some cases, existing data can be used for this purpose, and informative results can be obtained quickly. But often scientists will need to apply for funding to carry out full-fledged studies from scratch that attempt to improve on the initial study. Whereas laboratory scientists can carry out an experiment (or a series of experiments) in a matter of weeks or months to address a question and if it does not pan out, they can quickly shift to a new line of inquiry; the typical epidemiologic study involving a large population can take nearly a decade from the grant proposal stage to the publication of results. This reality has implications for the life of a putative hazard. If a hazard attracts the attention of scientists, a wave of reports based on more mature studies may appear a number of years following the initial report(s), and new studies may continue to appear for many more years. Sometimes, by the time the results of the later studies appear, the scientific community has all but lost interest in the question. In epidemiology, one is wedded to the data one has collected and has an obligation to publish results that might have been of burning interest ten years earlier but may now be old hat.

REGULATORY AGENCIES AND EXPERT COMMITTEES

When a putative hazard achieves a certain level of prominence, regulatory agencies are likely to become involved in order to demonstrate their responsiveness to a perceived threat. Here, at the interface of science and policy, scientists may be playing a political role as well as a scientific one, and the two roles may be hard to separate. Findings that would diminish the importance of the putative hazard or that would help to put it in perspective may be given short shrift. Certain kinds of evidence may be given more weight, and others may be downplayed in order to fit the conclusion that the committee wants to arrive at.[28] Studies that show the desired association may be rated as superior to those that fail to show an effect, in a travesty of impartiality. Finally, one can exaggerate what may be a small or borderline effect, making it seem more consequential than it is. By such emphases and omissions, one can make the evidence appear more uniform or meaningful than it is in reality. Regulators and health officials may overstate the evidence in order to promote a specific policy (reducing exposure to tobacco smoke) or to foster public consciousness of an alleged hazard (radon in homes, electromagnetic fields). Central to such instances in which hazards are inflated are a failure to be sufficiently critical of the quality of the available evidence and a disposition to accept the evidence as if it were more meaningful than it is. Scientists who are willing to apply a more critical standard and to say outright that a great deal of what is published on certain topics is of poor quality are in a distinct minority.[29]

In some instances, the very composition of an expert committee charged with writing a report—and particularly the leadership of the committee—can influence the conclusions and the tone of the report.[30] If a committee is heavily weighted with scientists who have a stake in a particular area, this can color the assessment of the hazard. Agencies, like individuals, can become wedded to a particular belief, and this can influence their evaluation of the evidence and even what projects are deemed worthy of funding.

POLITICIANS

When public concern about a perceived health hazard reaches a certain point, politicians may take it up to be responsive to their constituents. For example, in the late 1980s, concern about a danger from radon in homes became a focus of national attention, and legislators and federal agencies had a profound influence on how this scientific question was formulated and on the policies that were adopted. A number of senators and representatives took a vigorous role in addressing the suspected hazard. However, the congressional hearings held regarding radon tended to solicit expert testimony only from those scientists who saw domestic radon as a serious problem and not from equally qualified scientists who had compelling reasons to think otherwise.[31] Legislators also tended to favor the aggressive policies of the EPA, as opposed to the more conservative approach of the Department of Energy.[32] Similarly, in the early 1990s the Long Island senator Alfonse D'Amato became a spokesman for Long Island women who feared that environmental pollutants were contributing to what were thought to be high rates of breast cancer in their communities. Largely as a result of D'Amato's efforts, Congress allocated thirty million dollars to fund government-sponsored research to address the causes of breast cancer on Long Island.

While politicians can perform a vital role in advocating for increased funding for scientific research on important but neglected topics, their involvement in highly-charged issues can also contribute to the inflation of perceived health risks.

THE MEDIA

The most obvious and most proximate source of the hyping of risks is media coverage of stories involving health risks. Here there is a clear selection process at work since reports of positive findings—that is, of a hazard—are much more likely to be published and prominently featured than reports of no effect. The most conspicuous media reporting of suspected hazards is fundamentally antithetical to the essence of science. Science operates by the continual incremental accumulation of new data and the constant revision of theories to account for new and better data. In contrast, the "newsworthy" item is presented in isolation without the background or qualifications that would help put it in perspective.

Having said this, however, we need to distinguish between news reports that feature a potential threat on the one hand and broader and more critical articles discussing what is known on a particular topic on the other. I have been referring to the former, which give currency to a poorly substantiated hazard. I will cite many examples of the latter, which often do a superb job of explaining relevant considerations and putting a putative hazard in perspective. Sometimes an article combines both features, but in such cases one may have to read to the end to reach the critical commentary.[33]

The various groups and mechanisms discussed above do not exist in isolation but, rather, impinge on and influence each other. Scientists can respond to a public perception of a hazard and can both study a problem and produce findings that add to public concern. They can advocate for the need for further study of an issue when they may stand to benefit from special research programs. Regulatory and health agencies can respond to a perceived threat by setting aside special funding and by issuing reports and advisories. These special programs are then interpreted by the public and the media as confirming the existence of a hazard. It needs to be emphasized that a public aroused by anxiety can be useful to researchers, regulators, and politicians committed to pursuing a suspected hazard. After contributing to public concern regarding a specific hazard, scientists may then cite as a justification for further research the need to allay the public's fear. Some observers have noted that, as long as there is public concern and money to conduct studies, it is likely that scientists will continue to study an issue.

RISKS ARE SELECTED

I have already referred to the fact that certain types of exposures are viewed by the public as carrying a higher risk and as being less acceptable, in contrast to the way scientists view them. In addition, it is significant that not all agents have equal power to evoke the deep terror that is associated with certain types of hazards. For example, major known causes of chronic disease, such as smoking, heavy consumption of alcohol, obesity, and excessive exposure to solar radiation, do not engender the kind of anxiety that attaches to other perceived threats over which people have less direct control. This may provide a clue to one of the necessary conditions for a putative hazard to become a focus of societal concern. Hazards that are nondiscretionary and thus beyond our ability to control and that are invisible—such as electromagnetic fields, ionizing radiation, and chemical pollutants—have a greater potential to inspire terror than everyday, discretionary behaviors like smoking, drinking, weight gain, and exposure to the sun's rays.

Also, it is much easier for people to focus on external threats than to take responsibility for personal behaviors that are difficult to change but which have a much greater impact on health. Smoking and obesity are responsible for a large proportion of death and disability, and yet people find it very hard to change these behaviors—to quit smoking, change their eating habits, control their weight, and exercise more. Thus, when scientists, regulators, and the media overstate the significance of a potential environmental hazard (like electromagnetic fields from power lines, residential radon, or secondhand tobacco smoke), this plays into the public's predisposition to focus on certain types of risks. But it also happens that nondiscretionary exposures are often much harder to measure in scientific studies than personal factors, like smoking, alcohol intake, and body weight, and exposure in the general population tends to be at low levels. These realities add another critical element that characterizes such hazards, namely, uncertainty.

When examined in their own right, the distortions and exaggerations of health risks have much to tell us about the interplay between the science of public health and the wider society. The anthropologist Mary Douglas and the political scientist Aaron Wildavsky have argued that the selection of risks is "socially constructed" and gives expression to deep and often unconscious preoccupations of a society.[34] Douglas and Wildavsky identified fears concerning technology and environmental pollution as forming a powerful ideological complex that has a profound effect on the political process. More recently, the biochemist Bruce Ames has provided a vigorous critique of the tendency to assume that synthetic compounds must be harmful and that "natural" compounds must be good and the tendency to give undue weight to minute exposures to synthetic chemicals.[35] These kinds of unquestioned, deeply held beliefs can play a role in what scientific research gets funded and how it is interpreted by groups with a strong vested interest in a particular question, as well as by the public at large. Thus, there is a complex and inescapable, but largely unacknowledged, interplay between the parties and institutions that contribute to the manufacture or inflation of a hazard.

PARA-KNOWLEDGE

For all that distinguishes them from one another, the diverse parties that contribute to the manufacture of a hazard appear to have one thing in common. They are all susceptible to what has been called the "availability heuristic." This term, coined by the psychologists Daniel Kahneman and Amos Tversky, refers to the fact that our judgment regarding the likelihood, and, hence, the importance, of an event can be skewed by what our attention is focused on.[36] In essence, what is in front of us tends to blot out consideration of other factors, which may be more important. We are strongly influenced by the framing of an issue and by its vividness, or salience. Professional pollsters and psychologists are well aware of this phenomenon and have developed strategies for avoiding it. For example, if people are asked a single question about whether they think that exposure to "chemicals" causes breast cancer, a higher percentage will say yes than if "chemicals" is only one item in a list of possible causes. To give another example, our feelings about taking an airplane flight may be influenced by a recent airline crash involving hundreds of deaths, even though we know that in general flying is extremely safe. By the very act of focusing on a specific question and isolating it from its context, one is more apt to find evidence supporting its importance.

As a consequence of the kind of biases, vested interests, and narrow focus of attention described above, on certain questions a sort of *para-knowledge* can be built up and widely disseminated. By para-knowledge, I mean propositions that are given currency and are presented in such a way as to be beyond questioning. Instead of a rigorously impartial and comprehensive account of the relevant evidence, what we get is a selected account tailored for instrumental or partisan purposes. The propositions disseminated in this way become self-evident "facts" that "everybody knows." To give just a few examples: breast cancer is asserted to constitute an "epidemic" in certain regions of the country; radon and secondhand tobacco smoke exposure are claimed to each account for about one quarter of lung cancer cases occurring in never smokers; and exposure to secondhand smoke is officially stated to be responsible for about 50,000 deaths from heart disease each year and is labeled a cause of breast cancer in a report by a state environmental agency. This kind of factoid—detached from any sense of the evidence supporting the claim or any sense of how the alleged risk compares with established risks—is the end result of reports that overstate the evidence on a given question. Alongside of this instrumental para-knowledge, there are more rigorous and impartial assessments of the evidence. But, not surprisingly, these cannot compete with the much more compelling and satisfying—and often politically correct—claims of an effect. These articles may exist in the literature but will be rarely cited. Overall, outright criticism of the reigning consensus by professionals who are in a position to know is often quite feeble.

The distortion of information about health risks has consequences and costs—at once societal and individual—which are rarely acknowledged. These include tens of billions of dollars in remediation of what appears to have been a pseudo-problem (I am referring to electromagnetic fields from power lines). Clearly, resources devoted to a poorly-substantiated threat cannot be devoted to less sensational lines of work, which may prove to be more valuable. Another cost entails the needless anxiety about problems that were miniscule or non-existent, and the confusion about what are firmly established and important health risks. The amplification of hazards diverts attention from real and consequential dangers.[37] Perhaps most important is the corruption of science that occurs when, for reasons of political expediency and self-interest, responsible institutions and scientists attempt to create a consensus by ignoring divergent evidence. Not only are dissenting voices ignored or maligned, but the very undertaking is contrary to the essence of science, which relies on skepticism and an openness to new evidence and new interpretations.

***

I have tried to address this book both to the interested general reader and to specialists and students in public health. Chapter 2 is designed to give the nonspecialist "a feel" for what epidemiology is, the basic logic of epidemiologic inquiry, its wide range of application, and its value, when its findings are interpreted in a discriminating way. To prepare the reader for the case studies, I give examples of what substantial and well-established associations look like, in contrast to weak, ambiguous, and inconclusive associations. In view of the confusion caused by the seemingly unending series of news reports of studies often yielding contradictory results concerning threats, an attempt to make some of the central concepts of epidemiology accessible to a general audience seems like a worthwhile undertaking.

The core of the book (chapters 3–6) consists of four case studies of environmental hazards: environmental pollution and breast cancer; electromagnetic fields and a range of diseases; residential exposure to radon as a cause of lung cancer; and passive smoking as a cause of serious chronic disease. For each hazard, I trace the trajectory of the research studies over time and the ways in which factors internal and external to the science influenced the interpretation of the evidence and contributed to shaping the message that was conveyed to the public. Although it was necessary to discuss the nature of the evidence for each hazard, each of the case studies also tells a story that deserves wider attention. In order to recapitulate the main points of the chapters that follow, "major take-home points" are provided at the end of each chapter.

Each of the case studies has its own unique features. For one thing, the nature of the alleged hazard, its perception by the public, the circumstances surrounding exposure, and the available evidence is different in each case. For example, extensive evidence supports radon's ability to cause lung cancer, whereas decades of research have failed to yield any convincing evidence that exposure to low frequency electromagnetic fields in everyday life can cause disease. Also, the way in which external forces, including industry, advocacy groups, and government agencies impinged on the controversy differed in each case. Nevertheless, in spite of these differences, a number of common features are discernable. In all four cases, early evidence of a hazard based on limited or flawed studies was exaggerated, and actions taken by government agencies served to promote the existence of a hazard and to ignore evidence that would have helped present a more accurate picture.

Other controversies could have been included—such as those pertaining to cellular telephones and brain tumors or general air pollution and chronic disease, to name just two; however, the overall conclusions would have been essentially the same. One justification for my selection is that my four topics involve very different types of exposure: residues of chemical compounds in food and water (organochlorine compounds); electromagnetic fields produced by power lines and electric appliances; a naturally occurring element found in the earth that emits high-energy particles (radon); and indoor air pollution produced by a discretionary personal activity, namely, smoking. The very different nature of these hazards makes all the more striking certain parallels in the way the scientific evidence was interpreted and the hazard exaggerated.

Another reason for focusing on these particular hazards is that I have had a direct professional involvement in epidemiologic research on three of the four topics (passive smoking, electromagnetic fields, and risk factors for breast cancer). And while I have not conducted research on radon, I have long been involved in research documenting the effects of the foremost cause of lung cancer, cigarette smoking, as well as in the causes of lung cancer in those who have never smoked. Having done primary research and contributed to the literature on a number of these topics, I have an intimate acquaintance with the relevant bodies of evidence. In following the scientific work on these hazards over the years, I have tried to maintain the dual perspective of the practicing epidemiologist and the skeptical observer.

In addition to a close reading of the literature on my four topics, I have benefited from interviewing a number of colleagues who have done important work on some aspect of the questions I treat, as well as several advocates who have played an important role in making breast cancer a research priority. These interviews and follow-up discussions deepened my understanding of key issues and helped me to clarify both the facts and my argument. What was most instructive about these interactions was how open accomplished scientists and activists were in acknowledging the tenuous support for some of their hypotheses and the wrong paths we are all capable of going down. This kind of frank assessment added a different dimension to what one can glean from the published scientific literature.

2

EPIDEMIOLOGY
Its Uses, Strengths, and Limitations

The biggest promise of this method lies in relating diseases to the ways of living of different groups, and by doing so to unravel 'causes' of disease about which it is possible to do something.

—Jeremy N. Morris, 1955

Epidemiology appears forbidding and unapproachable to many people, starting with its association with "epidemic" and its erroneous association in some people's minds with "epidermis" and "entomology." The term derives from the Greek "epi" connoting "on" and "demos" meaning "people," thus conveying the notion of "on, or pertaining to, populations." There are many definitions, but they all have in common the basic idea that epidemiology is the study of the occurrence of health and disease in populations, with a view to illuminating the causes of disease and, ultimately, its control or prevention. Beyond its unwieldy name, epidemiology's technical terminology and statistical apparatus undoubtedly help to make it unapproachable to a lay audience. Add to this the fact that concepts like "risk," "association," and "causation" are notoriously slippery, and their technical meaning is often different from what gets communicated to the public. Finally, the many conflicting media reports of findings from epidemiologic studies create confusion as to what are truly important hazards that are worth paying attention to, and this state of affairs has affected perceptions of the discipline.[1]

All of this is unfortunate because the results of epidemiologic studies are potentially of enormous relevance to the general public and because the basic principles of epidemiology can be understood and appreciated by those lacking specialized training. It has been said that, when stripped of its technical and statistical apparatus, much of epidemiology is basically common sense—but an "uncommon common sense."[2]

It is not possible to give a comprehensive overview of epidemiology for the lay reader within the scope of a brief chapter.[3] My goal is more modest, that is, to explain a number of key concepts and principles of epidemiology and to give examples of risk factors that have been firmly established in contrast to potential hazards that have received a great deal of publicity but for which clear evidence of harm is lacking. This outline provides a framework for understanding the detailed case studies that form the core of this book, but it should also help more generally in interpreting reports of "hazards"—as well as claims of health benefits—in the media. By "interpreting," I do not mean evaluating the technical aspects of studies or reports but rather developing a "feel" for what epidemiology can achieve, as well as what its limitations are. Above all, I hope to give the reader a sense of perspective when confronted with dramatic claims concerning health effects.

THE BASIC SCIENCE OF PUBLIC HEALTH

Epidemiology is commonly referred to by its practitioners as "the basic science of public health." What it has to offer is, above all, a unique and powerful approach to questions having to do with health and disease. Based on groups and salient differences between groups, it involves a different way of thinking from the mindset of clinical medicine centered on diagnosis and treatment of the individual patient. Epidemiology takes a broad phenomenological view of illness and health in their existing context, incorporating relevant findings from a wide range of other disciplines, including microbiology, toxicology, genetics, industrial hygiene, sociology, pathology, nutrition, ecology, statistics, etc. Epidemiology occupies a privileged position in assessing health hazards, since, unlike experimental studies involving animals and cell culture, it focuses on the actual exposures of people as they are experienced in everyday life. Thus, agencies charged with evaluating the health effects of environmental agents give priority to epidemiologic data, when these are available. Both the strengths and the limitations of epidemiology stem from the same source—that is, from the fact that it uses observational data to attempt to unravel the complex causes of multifactorial, multistage chronic diseases in free-living populations.

Often, epidemiology takes as its point of departure a striking contrast, such as the fact, noted in the 1970s, that rates of hepatocellular carcinoma, the most common type of liver cancer, showed a striking worldwide variation, with the highest rates in East Asia and southern Africa and the lowest in the Americas, northern Europe, and Australia. This observation eventually led to the development of a vaccine against hepatitis B virus that confers protection against this major cause of liver cancer. Or the fact, noticed over three hundred years ago, that nuns had elevated rates of breast cancer. This observation led eventually to an understanding of the role of reproductive and hormonal factors in the development of breast cancer, with dramatic implications for its treatment and prevention. From such fertile starting points, and the collation of many other relevant facts, a picture is developed that points to specific hypotheses and lines of inquiry. The work

Box 2.1 A Partial List of Major Achievements of Epidemiology

Cholera and use of contaminated water supplies

Vitamin B (niacin) deficiency and pellagra

Smoking and cancers of the lung, mouth, larynx, esophagus, bladder, kidney, pancreas, and liver

Alcohol consumption and cancers of the mouth, larynx, esophagus, liver, and breast

Multiple risk factors for coronary heart disease: cholesterol, smoking, obesity, high blood pressure, and diabetes

Reproductive factors and cancers of the breast, endometrium, and ovary

Estrogen and endometrial cancer

Estrogen and progesterone and breast cancer

DES (diethylstilbestrol) and clear cell adenocarcinoma of the vulva

Exposure to coke oven emissions (in the steel industry) and lung cancer

Exposure to asbestos and lung cancer and mesothelioma

Folic acid deficiency and neural tube defects

Aspirin and Reye's disease

Lead and cognition

Hepatitis B virus (HBV) and primary cancer of the liver

Human papillomavirus (HPV) and cervical cancer

Aspirin and prevention of colorectal cancer

Oral contraceptives and prevention of ovarian cancer

Solar radiation and skin cancer

Fluoridation of water supplies and prevention of dental caries

Sleep position and sudden infant death syndrome

of pursuing these hypotheses, often carried out over decades, has led to stunning triumphs in public health affecting potentially millions of lives worldwide. Some of the major achievements of epidemiology are given in box 2.1.

One of the strengths of epidemiology is worth pointing out at the outset. This is that knowledge of an association permitting the identification of high-risk groups can, in some situations, make it possible to curtail the incidence of disease, even in the absence of an understanding of the mechanism of causation. In the legendary example of the London cholera epidemic of 1854, John Snow did not have to know that the organism that caused cholera was *vibrio cholerae*. He merely had to remove the handle of the Broad Street pump, thereby preventing the local residents from using a contaminated water source. Similarly, we do not know precisely which of the thousands of constituents of tobacco smoke are responsible for inducing lung cancer; nevertheless, knowledge of the association of smoking with lung cancer makes it theoretically possible to eliminate roughly 90 percent of lung cancers. Finally, the data are persuasive that if Americans reduced the prevalence of obesity, the overall death rate, as well as deaths from heart disease, diabetes, and certain cancers would be reduced.[4]

Epidemiology deals with probabilities that apply to individuals as members of a group rather than to individuals themselves. Its ability to predict events on the individual level is often quite limited. Even when we are dealing with what is agreed to be a strong association, like that of cigarette smoking and lung cancer, only a minority of smokers will develop lung cancer. Furthermore, among smokers who develop lung cancer, one cannot be certain that, in the case of an individual smoker, smoking was "the cause" since some people who have never smoked also develop the disease. Nevertheless, "on a population basis," the association between smoking and lung cancer is both strong and robust. The low predictive ability at the individual level of even strong epidemiologic findings is one source of misunderstanding of the value and use of such findings.

Because epidemiology deals with groups, the frequency of occurrence of a given disease has an important bearing on the ability to study it. This is especially the case when one is studying a factor which may make only a relatively subtle contribution to causing a given disease or when there are multiple factors which play a role in its causation, as is almost always the case. The more common the disease, the easier it is to obtain an adequate study population, or *sample size*, and sample size is one of the determinants of the statistical power to detect an effect of a hypothesized factor. As a general rule, the subtler the role a given factor plays in the causation of a disease and the poorer the accuracy with which one can measure that factor, the greater the sample size required. We have been able to learn a great deal about factors that contribute to heart disease, in part because heart disease is the leading cause of death in the United States and other Western countries. In contrast, studies of relatively rarer diseases, such as scleroderma and ALS (amyotrophic lateral sclerosis), are hampered by, among other things, the difficulty of accruing an adequate sample size.

Epidemiologists focus on *rates* of occurrence of disease because absolute numbers of cases are uninformative with regard to etiology unless they are related to the appropriate denominator, i.e., the population from which they arose. A rate is the frequency of some health event per population over a certain unit of time (usually one year). Rates can be presented for an entire population. Examples are the infant mortality rate (expressed in terms of the number of deaths in children less than one year old per 1,000 live births) or the breast cancer incidence rate (expressed as the number of new cases of breast cancer per 100,000 women). These figures are referred to as "crude" rates in that they are the average for an entire population. While useful for certain purposes, crude rates can mask large differences within specific subgroups that make up the population. Once one begins to look at the rates for specific subgroups (by age, sex, and race) within a population, one may begin to see striking contrasts, and these may point to hypotheses about the causes of a disease. Epidemiology deals with differences in the rates of disease according to the myriad factors—starting with age, sex, ethnicity, socioeconomic level, occupation, religion, smoking status, country of birth, etc.—which may lead to an understanding of etiology.

Fundamental to epidemiology, at the most basic level, is a comparison. Comparison is implicit in the definition of "epidemic," which is a sudden elevation in the number of cases of a disease *relative to the background, or baseline, level*. If comparison between two or more groups is fundamental to epidemiology, one needs to be sure that the comparison is a fair one. For this reason, the issues of confounding and bias are central concerns in epidemiology. Confounding refers to a muddying or obscuring of the association of the factor of interest with a disease by other factors that are correlated both with the factor of interest and with the disease. Confounding can lead either to a diminution of the observed effect of the factor of interest or to its inflation. A commonly used example of confounding involves comparing the health of residents of Alaska with that of residents of the lower states of the United States. If one were to compare the overall incidence of heart attacks among Alaskans to that among Americans in the lower forty-eight states, it would appear that Alaskans had much lower rates. However, the comparison would be misleading because it is confounded by a large difference in the age structure of the two populations being compared. The Alaskan population is considerably younger, and age is a risk factor for heart attack. When one makes the comparison taking age into account (i.e., when one compares "age-adjusted" rates), the difference in heart attack rates disappears.

To give another example, in assessing the possibility that coffee drinking increases the risk of bladder cancer, one would need to take smoking habits into account. This is because smoking is a known risk factor for bladder cancer and because coffee consumption and smoking are correlated. If one failed to take smoking habits into account in one's analysis, it is likely that coffee drinking would appear to be a risk factor simply due to its correlation with smoking.

A final, and more topical, example of confounding comes from studies of the effect of a woman's use of estrogen replacement therapy on her subsequent risk of coronary heart disease. For many years, observational studies were interpreted as indicating that women who had taken postmenopausal estrogens had lower rates of heart disease, and this finding figured prominently in the widespread promotion of estrogen replacement therapy by physicians in the 1980s and 1990s. However, it turned out that the finding was due to confounding, i.e., women who took estrogen replacement therapy (ERT) tended to be different, in a variety of ways, from women who did not. Specifically, ERT users tended to have a healthier lifestyle compared to nonusers. They tended to have lower body weight, were more likely to engage in physical activity, and were more likely to be nonsmokers and in general to be more health conscious—differences that would place them at lower risk of heart disease.[5] By the late 1990s and particularly in 2002, randomized controlled trials, which were less susceptible to confounding compared to observational studies (as will be discussed below), showed that, rather than protecting against heart disease, use of ERT actually increased its risk by a modest amount.

It should be noted that you can only adjust for a potential confounding factor if you have thought to measure it, i.e., include it in your study. Also, adjustment for a poorly measured confounder does not eliminate the possibility of *residual confounding*, since measurement of the confounder may not be adequate to fully capture its effect.

At the most basic level, epidemiology involves evaluating the association between a given characteristic, exposure, or "risk factor," on the one hand, and a state of health or disease, on the other. To give a few examples, using epidemiologic methods, researchers have studied the association between "handedness" (being left-handed versus being right-handed) and breast cancer; between seat belt use and injuries and deaths resulting from motor vehicle collisions; between the degree of elaboration of one's writing style, as disclosed by diaries kept by young nuns, and the subsequent likelihood of developing Alzheimer's disease; between exposure to television violence and rates of violent crime; between alcohol and illicit drug use and homicide; and between having a specific genetic makeup and the risk of virtually any disease. As these examples convey, the epidemiologic method can be used to study the influence of a wide range of social, behavioral, environmental, and genetic factors on health and disease. Many things that epidemiologists study are potential clues to the etiology of a disease, factors that it is hoped will lead to the primary contributors. In many cases, a risk factor is an indicator or marker of a phenomenon that may turn out, upon further study, to be important rather than the cause itself. This is another frequent source of misunderstanding.

The basic relation between an exposure and a disease can be represented schematically by the "two-by-two table" favored by epidemiologists. At the simplest level, both an exposure and a disease can be expressed in dichotomous terms (yes/no or present/absent). Either one has (or has had) the exposure or one has not, and either one has the disease (or one develops it within a specified period of time) or one does not. In the two-by-two table the letters a, b, c, and d represent the number of people in each of the four quadrants.

		Exposure	
		Yes	No
Disease	Yes	a	c
	No	b	d

To be sure, this is a simplified and idealized schema. In some situations, it may be difficult to find a group that is truly unexposed, and one often wants to look at different levels of exposure.[6] Nevertheless, this schema is extremely useful in describing the essentials of epidemiologic studies.

In a case-control study (described in the next section), the measure of association obtained from this table is called the *odds ratio*—because it is the ratio of two sets of odds: the odds of having the exposure if one has the disease divided by the odds of having the exposure if one does not have the disease. The odds ratio can be obtained by the formula a × d/b × c, which is referred to as the "cross-product ratio."

To give an example using actual data, let's fill in the results from a typical study relating cigarette smoking to lung cancer. Say we conducted a study in several hospitals that entailed interviewing all patients with a diagnosis of lung cancer and, as a comparison group, an equal number of patients with diseases unrelated to smoking (e.g., fractures, hernias, enlarged prostates). The resulting two-by-two table would look like this:

	Ever smoked	Never smoked	**Total**
Lung cancer	95	5	100
No lung cancer	50	50	100
Total	145	55	200

We see from the table that 95 percent of lung cancer patients were smokers, whereas only 50 percent of the comparison group were smokers. Using the cross-product ratio, the odds ratio = 95 × 50/5 × 50, which yields 19. In this study, a patient with a diagnosis of lung cancer was 19 times more likely to be a smoker than a patient with a diagnosis other than lung cancer. I will come back to the odds ratio and other measures of association shortly.

Since the objective of epidemiology is to detect and understand associations between a factor and a disease, the better the job of measuring the exposure of interest and the better the job of defining the disease entity in question, the better the chances of detecting a meaningful association. If measurement of the exposure is poor, this introduces "noise" and decreases the researcher's ability to detect an association. Similarly, if the disease in question cannot be precisely and objectively defined, it is unlikely that any associations are going to be biologically meaningful. Poor measurement of the exposure and the outcome is referred to as *misclassification*. This means that, due to poor measurement, some subjects who are actually exposed to the risk factor are mistakenly classified as "not exposed," and to the same extent those who are not exposed are classified as "exposed." In the same way, the absence of hard criteria for judging whether someone has the disease or not leads to classifying some people with the disease as "free of the disease," and conversely classifying some who are free of the disease as having the disease.

In contrast to random measurement error, which reduces one's ability to detect an association, *bias* refers to systematic errors or imbalances in the design or conduct of a study that can lead to distorted results. Common biases are described in the following section.

TYPES OF STUDY DESIGN

The different types of study design utilized in epidemiology are usually grouped under *descriptive* studies and *analytic* studies. Generally, descriptive studies appear early in the study of a disease and are followed by analytic studies, usually case-control and cohort studies. In the hierarchy presented here, there is progression toward a greater ability to identify causal factors. However, each type of study has its strengths and weaknesses and each can provide valuable evidence.

Descriptive Studies

Descriptive studies provide basic information about the occurrence of a disease in time and space, according to factors such as age, sex, ethnicity, social class, occupation, etc. Cases of a disease presenting at one or more hospitals or within a defined area or population are characterized as to demographic factors, lifestyle, behavioral factors, and environmental exposures. Such information is essential in gauging the magnitude of a public health problem and can suggest hypotheses that can be pursued using other study designs. But, because descriptive studies lack a control group, they cannot be used to assess causality directly. For example, studies describing the occurrence of a rare form of pneumonia and a rare form of cancer in young gay men in Los Angeles and New York City in 1981 provided the first clues to the etiology of what was later identified as AIDS. These studies revealed that men with these conditions had engaged in anal sex with many sexual partners and had also used amyl nitrite and other recreational drugs. At first, attention focused on amyl nitrite and its adverse effects on the immune system as an explanation for the disease, but further research implicated anal intercourse in the transmission of HIV.

Ecological Studies

Ecological studies, or correlational studies, correlate existing data on populations with existing data on disease rates. What distinguishes ecological studies from other types of studies is that they make use of data on exposure and disease pertaining to groups rather than individuals. An example of an ecologic study is the striking correlation on an international level, first noted in the 1970s, between per capita consumption of fat in the diet and death rates from breast cancer in women. Women in advanced "Western" countries, like the United Kingdom, Denmark, and the United States had both a high intake of fat and high rates of breast cancer, whereas countries like Japan and Thailand had a low intake of fat and low rates of breast cancer. Plotting fat intake against the breast cancer death rate for different countries yielded an impressive graph showing a more or less ascending trend in the breast cancer death rate with increasing fat intake, as shown in fig. 2.1.

Note that not every woman in the United States consumes the same level of fat in her diet nor has the same risk of developing breast cancer. Nevertheless, in an ecologic study, one uses only the *average* intake for the group and contrasts this with the average for other groups. Because they make use of aggregate data on population groups and because many other factors are likely to differ among the populations one is comparing, confounding is a major problem affecting ecologic studies, and these are useful mainly in providing clues that require more detailed investigation with other approaches. In the 1950s, when the first studies appeared showing an association between cigarette smoking and lung cancer, some commentators pointed out that, while there was a strong correlation internationally between per capita sales of cigarettes and lung cancer death rates in different countries, there was also a strong correlation between sales of silk stockings and lung cancer rates. Did this mean that silk stockings were a cause of lung cancer? Of course not. Rather, in more affluent countries, consumers could purchase both cigarettes and silk stockings. Both items reflect a certain level of economic development. So, the

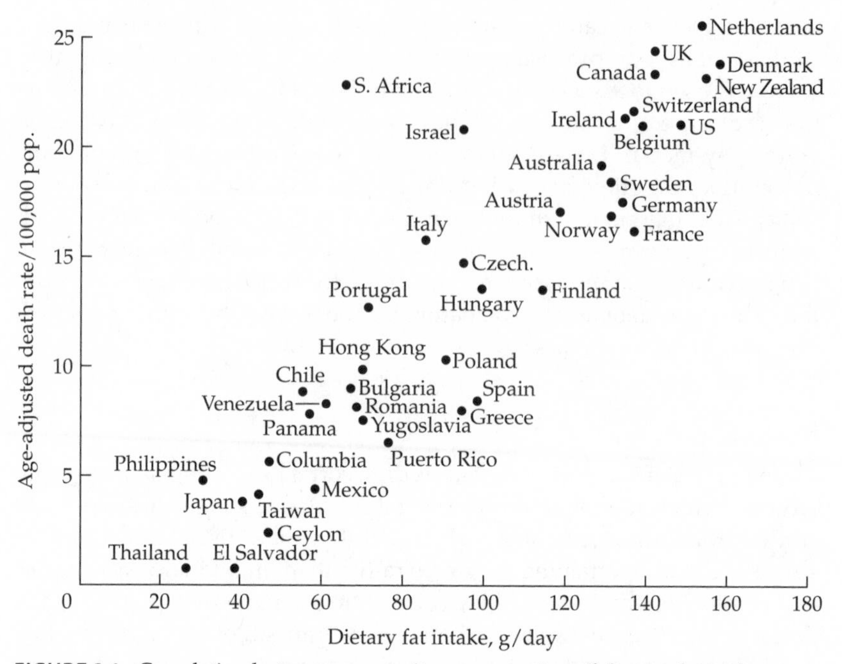

FIGURE 2.1 Correlation between per capita consumption of dietary fat and age-adjusted breast cancer mortality in forty-one countries. *Source*: Carroll and Khor, 1975.

existence of a strong ecologic correlation between an exposure and a disease is not a sufficient condition for any judgment regarding causality. Without additional and finer-grained evidence linking cigarette smoking with lung cancer, the association could simply be due to confounding by economic differences. When overly strong claims are made based on ecologic data, scientists refer to the "ecological fallacy," namely the unwarranted assumption that, just because a given factor is correlated with a disease based on aggregate data, this suggests (1) that the same correlation will hold at the individual level or (2) that the association is indicative of a causal relationship.

In spite of the pitfalls of ecologic data, it should be remembered that major triumphs in epidemiology and public health have started with observations based on ecologic data, such as the observation that prevalence rates of hepatitis B virus infection worldwide showed a striking correlation with rates of primary liver cancer.[7]

Case-Control Studies

The study of lung cancer using hospitalized patients referred to earlier is an example of a case-control study. In this type of analytic study, cases of a specific disease are identified through hospital records or disease registries, and a comparison group—controls—is selected by one of several methods. Controls should be drawn from the base popluation from which the cases arose and should be at risk of the disease under study. Most commonly, either hospital controls or population controls are used. The former could be patients hospitalized for conditions other than the disease of interest. Population controls can be selected from the community using random digit dialing, Medicare rosters, motor vehicle registry lists, or other sources. Information on exposure can include information recorded in the hospital record (when hospital controls are used), information collected in personal interviews, usually using a structured questionnaire, or analyses carried out on biological specimens obtained from cases and controls. Typically, in case-control studies involving a personal interview, participants are asked detailed questions about a range of relevant factors, including sociodemographic characteristics, occupation, lifestyle factors (smoking, alcohol consumption, diet), menstrual and reproductive factors (in women), medical history and use of medications, physical activity, and environmental exposures. In addition to interview data and biological specimens obtained from participants, measurements can be made in the current and previously occupied homes, as was done in studies of radon and electromagnetic fields.

The analysis of a case-control study entails a comparison of the level of exposure to the factor of interest among cases to that among controls. As we saw above, the association is expressed by calculating the odds ratio summarizing the two-by-two table or a more elaborate table classifying subjects as to their level of exposure. In addition, one needs to determine whether cases and controls differ on other factors, which might account for any observed difference in the study factor. If a third factor is related both to the exposure of interest and to the disease, it is a potential confounder and needs to be taken into account. This means that one attempts to collect information on a variety of factors that are known or potential confounders.

Case-control studies have a number of advantages. They enable the researcher to enroll large numbers of what may be a rare disease by including multiple hospitals, and they are relatively inexpensive to conduct. Additionally, they can yield results in a relatively short time. However, case-control studies are susceptible to two types of systematic bias that can distort the results. First, *selection bias* can occur if either cases or controls are not representative of the population from which they are drawn with respect to the exposure of interest. This could arise due to a low response rate among cases or controls or due to the use of a different mechanism for enrolling cases and controls. For example, in a study of electromagnetic fields and childhood cancer, cases were identified through a cancer registry, whereas controls were identified through random digit dialing. Thus, only households with telephones were eligible for control selection. It was pointed out that this difference may have introduced selection bias due to a higher socioeconomic status of controls compared to cases. The second type of bias is referred to as *information bias* (or *reporting bias*) and stems from the fact that one is obtaining information about exposure after the disease has been diagnosed. Cases may answer questions differently from controls because people who have recently been diagnosed with a serious illness tend to ruminate about what could have caused their condition. For example, if one is interested in collecting information about exposure to use of hair dyes in a case-control study of breast cancer, cases may tend to go back over their history more carefully and report greater exposure simply due to their motivation to explain their condition. Controls do not have the same motivation. Furthermore, collecting information by interview relies on memory, and in a case-control study one may be interested in information relating to events in the distant past.

Cohort Studies

Cohort studies can be viewed as the inverse of case-control studies. That is to say, rather than starting with the disease of interest, one starts with the exposure of interest. Thus, unlike the case-control study, in which one is looking backward in order to identify the cause of the disease, the cohort study follows the actual temporal sequence, starting with exposure and leading to the occurrence of disease. Cohort studies are also commonly referred to as "prospective" or "longitudinal" studies. Typically, one assembles a cohort of healthy individuals at the outset and assesses their exposure to factors of interest and then follows these individuals for a number of years in order to monitor their health status and the occurrence of "outcomes" of interest. For example, in 1959 the American Cancer Society enrolled over one million men and women by means of volunteers who contacted neighbors and friends with whom they expected to maintain contact over the coming years. In order to be eligible, a household had to have a person between the ages of 40 and 79. Participants completed and returned a four-page questionnaire including questions on smoking habits, alcohol consumption, exposure to dusts and chemicals, height and weight, how many hours of sleep they averaged per night, and their current state of health. The cohort was followed until 1972 and cause of death among its members was determined from death certificates. This early cohort study was a source of many important findings, including those documenting the link between smoking and heart disease and lung cancer.

The measure of association derived from a cohort study is the *relative risk*, which is simply the rate of occurrence of the disease of interest in those with the exposure divided by its rate of occurrence in those without the exposure. This corresponds to the odds ratio in a case-control study.

Compared to the case-control study, the principle strength of the cohort study is that it avoids selection bias and information bias. As long as follow-up of the cohort is largely complete, selection bias is not a threat, and since information on exposure is collected prior to the onset of disease, information bias is not a problem. The principle limitations of the cohort design are that cohort studies are expensive and can require a considerable amount of time to produce results. However, once established, cohort studies tend to become extremely valuable resources. A number of cohort studies have been in progress for 40–50 years and have provided a wealth of information as the participants have aged. Biological specimens that were stored can be used to test new hypotheses that come along.

Cohort studies can be carried out in the general population (American Cancer Society, the Framingham Study), occupational groups (uranium miners, steelworkers, agricultural workers, etc.), religious groups (Mormons, Seventh Day Adventists), and professional groups (doctors, nurses, teachers) or by enrolling the members of other organizations (Kaiser-Permanente health maintenance organization and the AARP).

Completeness of follow-up of the original cohort is the most crucial factor affecting the validity of cohort studies. If a sizeable proportion of the participants is "lost to follow-up" and the reason for loss to follow-up is related to the exposure under study, this could bias the results.

Randomized Controlled Trials

Case-control and cohort studies are observational in nature in that information is collected on free-living individuals, and one attempts to measure exposure to the factor of interest as best one can. In contrast, a randomized controlled trial is an *experimental* design. This type of study does not lend itself to assessing the association of environmental exposures with disease, which concerns us here. However, an understanding of the randomized controlled trial will help us understand some of the limitations of observational studies. In a randomized controlled trial, individuals who meet certain criteria and agree to participate are randomly assigned to different exposure groups. The design is similar to that of the standard animal experiment in which genetically identical animals are divided into two groups, differing only in the fact that the "experimental group" is exposed to the agent or substance under study, while the "control group" is not. (In practice, there can be more than two groups, depending on the number of dosage levels or different agents being studied.)

Suppose that one wanted to test the hypothesis that eating a high-fiber diet was protective against developing polyps of the colon, which can go on to develop into colon cancer. One would likely enroll a cohort of men and women of an age where these polyps occur fairly frequently, say, in their fifties and sixties. One would then randomly assign participants to one of two groups, referred to as the "intervention group" and the "control group." The intervention group would be instructed to eat a high-fiber diet, and the controls would be instructed to eat their usual diet. Both groups would undergo screening for polyps at regular intervals. Due to their having been randomly allocated, individuals in the two groups should be similar in all other respects except for their assignment either to a high-fiber diet or to their usual diet. Because the two groups are balanced in terms of a wide range of factors that could affect their risk of developing polyps in the colon (age, smoking, alcohol consumption, weight, physical activity, past diet, etc.), this experimental design enables one to isolate the effect of the exposure of interest, in this case, fiber. Because of their experimental design, randomized controlled trials are referred to as the "gold standard" in medicine and epidemiology. Note that by balancing the two or more groups in terms of extraneous factors by randomization, the randomized controlled trial avoids the problem of confounding, which is so fundamental in observational studies.

Randomized controlled trials can only be used in certain situations. Since it would be unethical to subject human subjects to an exposure known or likely to be harmful, one can only test exposures (or treatments) that are thought to have potential benefits or relative benefits. Thus, one could not conduct a randomized controlled trial in which one assigned nonsmokers to smoke cigarettes of widely different tar levels in order to learn about their long-term health impact. One could, however, conduct a randomized trial among smokers to compare the effects of different methods of smoking cessation. In spite of the increased rigor of the experimental design, randomized controlled trials do not always provide a definitive answer to a question. In the example given above of allocating healthy men and women to be on either a high-fiber diet or to continue eating their usual diet, if the study finds no effect on the likelihood of developing polyps, this may be because the level of fiber prescribed to the intervention group was inadequate; the study period was too short; one might have to consume a high-fiber diet starting much earlier in life for fiber to protect against polyps; or other reasons. In other words, in a given study one can only test specific aspects of a question. Thus, it is important to realize that depending on the specific question under study, different types of studies can provide valuable evidence for or against an association. Each type of study design has its strengths and limitations. This is why when there is a convergence of supporting evidence from different types of studies, this greatly strengthens the case for a given factor playing a causal role.

MEASURES OF ASSOCIATION

If there is no association between an exposure and a disease, in the two-by-two table the proportions of those with the exposure would not differ between those with the disease and those without the disease, and the relative risk would be equal to 1.0. This is referred to as the "null value." If a study has an adequate sample size and the relative risk is above 1.0, and especially if it is well above 1.0, then the exposure of interest shows a positive association with the disease, and it may be a risk factor for the disease. If the relative risk is below 1.0, say 0.5, for example, then the factor under study shows an "inverse association" with the disease and may be "protective." In order to account for the possibility that the observed association could be due to chance, one conventionally provides a *95 percent confidence interval* bracketing the calculated odds ratio or relative risk. If the confidence interval does not include the null value (1.0), then the odds ratio or relative risk is said to be *statistically significant*. What this means is that the result is unlikely to be due to chance. More precisely, if one repeated the same study one hundred times, in ninety-five of those repetitions one would obtain a value for the odds ratio lying between the upper and lower bound of the confidence interval. Conventionally, one accepts a 5 percent possibility that the observed association could be due to chance. In the two-by-two table based on the hypothetical study of lung cancer, described earlier, the 95 percent confidence interval associated with the odds ratio of 19.0 was 8.0–43.0. Although, one can never completely rule out chance as a factor in any particular study, an association like that of smoking in this hypothetical study is unlikely to be due to chance.

A word about "statistical significance" is in order. Statistical significance is easily abused, and some journals discourage undue emphasis on it. Not every finding that is statistically significant is biologically significant, but too often the fact that a given result is statistically significant is used to imply precisely this. More relevant to gauging the importance of a finding is the consistency with existing evidence from other sources and the impact on a population basis. Most important discoveries in the area of epidemiology do not depend primarily on tests of statistical significance to demonstrate their importance.

EXAMPLES OF ESTABLISHED ASSOCIATIONS

We are now in a position to look at some examples of findings from different types of studies in order to understand the difference between established associations and associations where the evidence is weak, inconsistent, and inconclusive. Table 2.1 shows the findings from epidemiologic studies in which the presumptive factor has been firmly established to play a role in the disease based on research conducted over many years. (The relative risks in tables 2.1–2.3 are selected from individual studies or, in some cases, meta-analyses, on a given topic. They should not be viewed as definitive but, rather, as an indication of the strength of the association. The magnitude of the relative risk for smoking and lung cancer, for example, will vary from one study to another.) Note that these examples were

TABLE 2.1 Relative Risks for Different Diseases Associated with Exposure to Different Risk Factors

Risk Factor	Disease	Relative Risk	95% CI[a]
HPV (Human papillomavirus)	Cervical cancer[b, 8]		
HPV negative		1.0 (reference[c])	—
HPV positive		90	71–114
HBV (hepatitis B virus)	Liver cancer[9]		
HBV negative		1.0 (reference)	—
HBV positive		98	na
Cigarette smoking	Lung cancer (males)[10]		
Never smoker		1.0 (reference)	—
Former smoker		3.5	2.8–4.4
Current smoker		11.9	9.6–14.7
Alcohol consumption	Cancer of mouth[11]		
Nondrinker		1.0 (reference)	—
Heavy drinker		4.5	na
Family history of breast cancer	Breast cancer[12]		
No family history		1.0 (reference)	—
One or more first-degree relatives with breast cancer		2.5	1.5–4.2
Age at first birth	Breast cancer[13]		
Before age 20		1.0 (reference)	—
35 and above		2.3	na
Active smoking	Heart disease[14]		
Never smoker		1.0 (reference)	—
Former smoker		1.2	1.1–1.3
Current smoker		1.5	1.5–1.6
Hormone replacement therapy	Breast cancer[15]		
Never used		1.0 (reference)	—
Used >5 years		1.2	1.1–1.4
Silicone breast implants			
No	Connective tissue disorders[16]	1.0 (reference)	—
Yes		1.0	na

[a] CI, confidence interval; na, not available.

[b] Superscript numbers indicate note reference.

[c] Reference risk category (1.0) for a person with no or minimal exposure.

selected for illustrative purposes, and there are many other examples that could be cited. In the first two entries in table 2.1, exposure was assessed by means of a clinical assay performed on blood to measure levels of the human papillomavirus (HPV) or the hepatitis B virus (HBV) antigen. In the subsequent entries, exposure information was obtained by asking the subjects questions about smoking, alcohol consumption, hormone replacement therapy, etc. The relative risks are shown in decreasing order, ranging from 90 to about 1.2. (For the sake of comparison, the bottom line of table 2.1 shows the results of a large study of silicone breast implants and connective tissue disorders, where there is no association, i.e., the relative risk equals 1.0.)

When one is dealing with a relative risk of 90 or 100 in a well-designed study, it is unlikely that the result is due to confounding or bias. In the case of smoking and drinking, the magnitude of the association is still substantial and has been consistently reported in many studies. Even though the value of the relative risk is much smaller when it comes to the association of a number of breast cancer risk factors, these characteristics have been consistently shown to be risk factors for the disease.

Table 2.2 shows representative results from studies investigating a number of environmental exposures, including secondhand tobacco smoke, air pollution, electromagnetic fields, DDT exposure, and residential radon exposure, some of which will be discussed in the extended case studies in the following chapters. Here one is dealing with excess risks of a small magnitude, below 2.0. But, in addition to the small magnitude of the apparent increase, another crucial aspect of these studies is the difficulty of assessing the exposure in question. In the examples in table 2.1, information on exposure status of the study subjects was either highly accurate (a highly specific clinical assay) or reasonably accurate, even if obtained by self-report.

TABLE 2.2 Findings from Epidemiological Studies of Selected Environmental Exposures

Risk Factor	Disease	Relative Risk	95% CI
Exposure to a spouse:	Lung cancer in lifetime nonsmokers[a, 17]		
Spouse never smoked		1.0 (reference)	—
Spouse smoked		1.24	1.13–1.36
Exposure to indoor radon	Lung cancer in nonsmokers[18]		
Low exposure		1.0 (reference)	—
Heavy exposure		1.2	0.4–3.1
Exposure to EMF	Childhood leukemia[19]		
Low exposure		1.0 (reference)	—
Heavy exposure		1.7	1.2–2.3
Air pollution	Lung cancer[20]		
Least polluted		1.0 (reference)	—
Most polluted		1.03	0.80–1.33
DDT level	Breast cancer[21]		
Lowest fifth		1.0 (reference)	—
Highest fifth		0.82	0.49–1.37

[a] Superscript numbers indicate note reference.

For example, even though people may not remember the details precisely or may underreport certain exposures, information regarding smoking habits, alcohol consumption, estrogen replacement therapy, family history of breast cancer in first degree relatives, and age at first live birth and number of full-term pregnancies is reasonably accurate. While imperfect, the data permit one to contrast those with relatively heavy to those with low, or no, exposure. In the case of silicone breast implants, exposure information based on surgical records achieves a high degree of accuracy.

In contrast, for the exposures shown in table 2.2, exposure assessment is much more problematic and much weaker. In the case of secondhand tobacco smoke exposure, typically, one can obtain information about whether the subject's spouse smoked, how long the subject was married, how much the spouse smokes (smoked), whether other members of the household smoke, and whether, how often, and how long he/she was exposed to smoke at work and in other venues. However, this is far from having a quantitative measure of the amount of smoke a nonsmoker has been exposed to over the years. For residential radon exposure, typically, one can obtain a residential history and take year-long measurements in several rooms of the current and previous homes occupied for a substantial number of years. But here, again, the results are very far from providing a quantitative estimate of lifetime exposure as one can obtain—even if in an admittedly crude way—for lifetime cigarette consumption, as approximated by "pack-years" of smoking (obtained by multiplying the average number of packs of cigarettes smoked per day by a smoker by the total number of years he or she smoked). Similar qualifications apply to estimating a person's lifetime exposure to DDT, electromagnetic fields, or air pollution.

Thus, a crucial aspect of epidemiologic studies providing information on human risks is the quality of the exposure information that can be obtained. It may be that if one could do a better job of measuring the exposure, one might see a much stronger association, or, alternatively, that one might see no effect. But, in the absence of a technical breakthrough—in the form of the development of a biological marker that provides a reliable indicator of lifetime exposure, as does pack-years of smoking or serologic evidence of infection—things are likely to remain frustratingly uncertain due to the inadequacy of the assessment of exposure and the extremely weak nature of the observed associations.

When concerns arose in the early 1990s that silicone breast implants were causing a variety of diseases in women, numerous studies were carried out using different designs to assess the claims of adverse health effects ranging from ill-defined connective tissue disorders and rheumatoid arthritis to breast cancer. Because "the exposure," i.e., having had silicone breast implants, was clearly defined and well documented in medical records, the resulting studies were of a high caliber. These studies uniformly showed no evidence that women who had had breast augmentation had any higher incidence of the suspected conditions than women who had not had the procedure.[22] As a result, millions of women who had had silicone breast implants had their anxiety about their health allayed. Thus, when one is dealing with a well-defined and well-documented exposure, epidemiologic research can achieve quite clear-cut results.

If the estimate of the relative risk of disease due to an exposure is uncertain, any estimate of the impact of the exposure on the general population is going to be uncertain, because calculation of the proportion of disease attributable to an exposure incorporates the relative risk. This is why some observers have questioned estimates of the numbers of lung cancer and heart disease deaths in never smokers (3,000 and 35,000–53,000, respectively) attributed to secondhand tobacco smoke exposure and the number of lung cancer deaths attributed to residential radon exposure (up to 20,000).[23]

JUDGING WHETHER AN ASSOCIATION IS CAUSAL

To say that exposure x is "associated" with disease y is simply to say that the two variables are correlated. Although the existence of an association between two phenomena is a necessary condition for the existence of a causal association, it is axiomatic that association does not prove causation. When the first studies reporting an association between cigarette smoking and lung cancer appeared in the early 1950s, the science of epidemiology, particularly with regard to chronic diseases with long latency periods, was still in its infancy. Some eminent statisticians were skeptical that case-control and cohort studies could provide an unbiased assessment of the association since these types of studies did not conform to the ideal of experimental design in which subjects are randomly assigned to the intervention or control group. Furthermore, when dealing with chronic diseases, such as heart disease or cancer, one cannot expect the same kind of rigorous proof as that provided by Koch's postulates for acute infectious diseases. This is because chronic diseases have long latency periods and complex, multifactorial etiologies.

Judgments about the causality of complex, multifactorial diseases are more problematic than our notions derived from our early life experience allow for.[24] Some observers have pointed out that all diseases have a number of "component causes," all of which are necessary, as opposed to a single sufficient cause.[25] This reminds us that our knowledge of the causation of these diseases is always incomplete and that other contributing factors remain to be identified. Further, it has been argued that there is no clear set of criteria that can be used for judging whether a given association is causal.[26,27] Nevertheless, findings like those summarized in table 2.1 represent knowledge that has been repeatedly confirmed and buttressed by new research findings and has been the basis for interventions that have proven to reduce the incidence of specific diseases. Thus, in practical terms, table 2.1 shows causal associations.

In the mid-1960s the British statistician Austin Bradford Hill proposed a set of considerations for judging the causality of an association,[28] and these were included in the first *Report of the U.S. Surgeon General on Smoking and Health*.[29] Although Hill identified nine criteria, four are generally considered the most important: the magnitude of the association, consistency, temporal relationship, and coherence of explanation. Hill argued that the greater the *magnitude of an observed association*, the more likely it is to be causal. One reason for this is that the greater the magnitude of an association, the less likely it is to be due to confounding by some extraneous factor or the play of chance. *Consistency* refers to the fact that the association should be observable in different studies carried out in different populations. *Temporality* refers to the logical necessity that the exposure (cause) must precede the disease (effect). Finally, *coherence of explanation* alludes to the requirement that all relevant information concerning exposure and occurrence of the disease in question should be mutually reinforcing. This is often referred to as "biological plausibility." One additional criterion that is frequently added to Hill's original criteria is that of a *dose-response relationship*. That is, the greater the exposure (dose), the greater the probability of developing the disease (response). The demonstration of a dose-response relationship between a risk factor and a disease adds a crucial element to the evidence for a causal association. Table 2.3 shows the dose-response relationships between the number of cigarettes smoked per day and the risk for lung cancer and heart disease, between alcohol intake and cancer of the mouth, and between number of years of using hormone replacement therapy and risk of endometrial cancer.

As a number of critics have recently pointed out, there are exceptions to all of Hill's criteria.[30,31] For example, as shown in table 2.1, the relative risk does not have to be large when one is dealing with a causal association, and the requirement of a dose-response relationship is not always met. Beyond any fixed set of criteria or guidelines, it can be agreed that for any given question the relevant evidence needs to be evaluated critically on its own terms. Nevertheless, although they may not be uniformly applicable in all cases, Hill's criteria are still considered to have great value in assessing the evidence on a given question.[32,33]

Let's briefly summarize what is known about cigarette smoking as a cause of lung cancer using Hill's criteria for judging causality. First, the magnitude of the association is substantial: current smokers (typically those who have smoked for three or four decades) have about a 20-fold increased risk of lung cancer in men and a somewhat lower relative risk in women. Second, regarding consistency, hundreds of studies carried out in different countries and different groups within countries and in both sexes show similar results. Third, typically smokers begin smoking in their teenage years and, unless they quit, smoke for many decades before the average onset of lung cancer (about age 62)—thus, the exposure/cause clearly precedes the disease/effect. Fourth, as shown in table 2.3, smoking shows a classic dose-

TABLE 2.3 Relative Risks for Different Diseases by Level of Exposure to Different Risk Factors

Risk Factor	Disease	Relative Risk	95% CI
Cigarette smoking	Lung cancer (males)[a, 34]		
Never smoker		1.0 (reference)	—
Former smoker		3.5	2.8–4.4
Current smokers of:			
1–9 cpd[b]		4.1	2.9–5.8
10–19 cpd		7.9	6.1–10.1
20 cpd		12.5	10.0–15.6
21–39 cpd		16.4	13.0–20.8
40–80 cpd		18.7	14.5–24.0
Alcohol consumption	Cancer of mouth[35]		
Less than 1 oz/week		1.0 (reference)	—
1–4 oz/week		1.1	na[c]
5–9 oz/week		1.4	
10–20 oz/week		1.8	
20+ oz/week		4.5[d]	
Hormone replacement therapy	Endometrial cancer[36]		
Never used		1.0 (reference)	—
<1 year		1.4	1.0–1.8
1–5 years		2.8	2.3–3.5
5–10 years		5.9	4.7–7.5
>10 years		9.5	7.4–12.3
Active smoking	Heart disease[37]		
Never smoker		1.0 (reference)	—
Former smoker		1.2	1.1–1.3
Current smokers of:			
1–9 cpd		1.2	1.1–1.3
10–19 cpd		1.4	1.3–1.5
20 cpd		1.6	1.5–1.7
21–39 cpd		1.8	1.6–1.9
40+ cpd		1.9	1.7–2.1

[a] Superscript numbers indicate note reference.
[b] cpd, number of cigarettes smoked per day.
[c] na, not available.
[d] Relative risks are adjusted for smoking and other risk factors.

response relationship with lung cancer. In the example given, as the average number of cigarettes smoked increases, the relative risk of developing lung cancer increases correspondingly, reaching a relative risk of about nineteen in male smokers of forty or more cigarettes per day. In addition, an early age of starting smoking age and greater number of years of smoking also make independent contributions to increasing risk. There are other pieces of information that further strengthen the case for a causal role of smoking in the development of lung cancer. These include the fact that when the exposure is removed, i.e., when a smoker gives up smoking and remains abstinent, the risk decreases over time approaching (but not necessarily ever reaching) that of someone who never smoked. Additionally, known carcinogens—such as benzo-a-pyrene—have been isolated from tobacco smoke, and smokers have evidence of genetic damage that puts them at increased risk of lung cancer. Finally, mice whose skin was painted with tobacco smoke condensate developed tumors at the site of application. Taken together, these findings amount to a varied and robust case for smoking playing a causal role in the development of lung cancer. Smoking is the most studied cause of cancer. Even the tobacco industry, which long claimed that the evidence was "only statistical" in nature, has come to accept it.

Even though the association of hormone replacement therapy (early formulations used only estrogen) with breast cancer shows only a modest relative risk, this has been confirmed in many studies, and this finding is supported by many other types of evidence.[38] These include the association of reproductive and menstrual factors with risk of breast cancer, which suggest that the greater a woman's exposure to endogenous estrogen, the greater her risk of breast cancer. Most recently, strong support for a role of estrogen in breast cancer has come from treatment of breast cancer with aromatase inhibitors, which block the formation of estrogen in adipose tissue, the major source of estrogen in postmenopausal women. Thus, even though the relative risk associated with hormone replacement therapy is quite small—on the order of 1.3—based on the consistency in the results of many epidemiological studies and the coherence with what is known from clinical studies, the association of estrogen with breast cancer is reasonably firmly established.

INTERACTION

The effect of a given exposure may not be uniform across an entire population but may be greater in some subgroup, such as the elderly, smokers, or those with certain medical conditions. Another way to describe this phenomenon is that the exposure of interest may be either *enhanced* by or dependent on a second exposure or characteristic. This is commonly referred to as interaction. A classic example of interaction is the fact that individuals who both smoke and consume alcohol are at increased risk of cancers of the mouth compared to those who only smoke or only drink. Table 2.4 shows the interaction between smoking and alcohol consumption in the development of cancer of the mouth and pharynx. Here one sees an increase in risk with increased smoking among those who do not drink (top row), as well as an increase in risk with increasing alcohol consumption among nonsmokers (first column). In fact, within each level of one risk factor, there is an increase in risk with increasing level of the other factor. But

TABLE 2.4 Relative Risk of Oral Cancer According to Level of Exposure to Smoking and Alcohol

		Smoking (cigarette equivalents/day)			
		0	<20	20–39	40+
Alcohol (oz/day)	0	1.00	1.52	1.43	2.43
	<0.4	1.40	1.67	3.18	3.25
	0.4–1.5	1.60	4.36	4.46	8.21
	1.6+	2.33	4.13	9.59	15.5

Risks are expressed relative to a risk of 1.00 for persons who neither smoked nor drank. *Source*: Rothman and Keller, 1972.

the greatest increase in risk occurs among those who are both heavy smokers and heavy drinkers. This group has a 15-fold increased risk compared to those who neither smoked nor drank. Other examples of interaction are the fact that exposure to either asbestos or radon have a much greater effect in smokers than in nonsmokers. The issue of interaction will be central when we consider the effects of radon exposure (chapter 5). Evidence that the effect of a given exposure is greatest in, or limited to, a particular group can have important implications for identifying those at highest risk, designing an appropriate intervention strategy, and formulating policy.

Since we are always dealing with multiple causes, attention to interaction between two different factors in producing a disease will receive increasing attention in the future. Thus, an understanding of why some smokers develop lung cancer and other diseases, while others do not, entails an understanding of the interaction between smoking and genetic makeup in terms of the enzymes that play a role in the activation or detoxification of the carcinogens and other toxins in tobacco smoke. In the era of the Human Genome Project, the question of the interplay between environmental exposures and lifestyle factors, on the one hand, and genetic makeup, on the other, has emerged as a major area of study.

META-ANALYSIS

Often there are a number of small studies addressing a question and providing inconsistent results. In this situation, epidemiologists have resorted to the technique of meta-analysis to combine the results from the individual studies in order to obtain a summary estimate that is more stable due to the larger sample size. Meta-analysis can be thought of as a process of taking a weighted average of the results of a number of individual studies. The technique was originally developed to combine data from a number of randomized controlled trials to achieve a more precise measure of the effect of a treatment. In this case, the combining of studies can be justified on the basis that each trial provided an unbiased estimate of the effect of treatment, owing to the experimental design.[39] When it comes to observational studies, the results of different studies can be affected in different ways by biases and confounding, and the assumption that their results can be arithmetically averaged is questionable.[40,41] Also, often the individual studies were not designed using common definitions and criteria. For all of these reasons, one may be combining results that are not comparable. Furthermore, combining studies that all suffer from the same biases does nothing to improve the validity of results.[42] Another potential problem is that meta-analysis depends on the published literature, and, if certain studies have not been published, this could skew the results of the meta-analysis. Some have suggested that, rather than conducting a meta-analysis of observational studies, more can be learned from examining the reasons that different studies yield different results. In other words, meta-analysis is no substitute for careful examination of the design and results of the individual studies in an attempt to determine whether individually and collectively they supply convincing evidence for an association. Thus, the uncritical use of a meta-analysis to summarize the results from observational studies of difficult to measure exposures can be misleading.

POOLING OF STUDIES

In contrast to meta-analysis, which is carried out by extracting the key results from published papers, pooling of data from a number of independent studies involves the collaboration of researchers who undertake to reanalyze the data from individual studies using a common strategy. While this permits greater uniformity and comparability to be achieved, there may still be aspects of the different studies that are not comparable. For example, different studies may have used different definitions of who qualifies as a never smoker (as is true of pooled radon studies) or may have used different measurement techniques. There may also be other differences between the studies being combined. For example, the effect of radon in a study carried out in northern China, where indoor air pollution from cooking and heating is a problem, may be different from its effect in Western countries.

Pooling of comparable studies to enable a more precise estimate of the effect of a given exposure, as well as exploring the robustness of the data, is a valuable tool, especially when one is dealing with a weak or subtle effect or attempting to detect an effect in a particular subgroup. However, when more and more studies are pooled and a clear signal does not emerge, this conveys important information that needs to be acknowledged.

ABSOLUTE RISK VERSUS RELATIVE RISK

Until now we have focused on the relative risk (or the odds ratio) and its use in evaluating the causality of an association. The relative risk is the most commonly used measure of association in epidemiologic studies, and this is what gets reported most often to the public. This is understandable, given that the relative risk has a straightforward interpretation: the risk in those exposed to a given agent is, say, two or three times greater (or less) than the risk in those who are not exposed. However, while useful in assessing etiology, the relative risk conveys no information about the impact of a given risk factor on the general population. This is because the relative risk is independent of the prevalence (i.e., the frequency) of the risk factor and the incidence of the disease in the population. A relative risk of 10.0 can result from differing ratios 10:1, 1,000:100, or 1,000,000:100,000.[43]

Unlike the relative risk, the *absolute risk* describes the population impact of a given risk factor and indicates how many cases of a disease would be eliminated if the risk factor were removed. Thus, the absolute risk, or risk difference, is the incidence of the disease of interest among the exposed minus its incidence among the nonexposed. The absolute risks corresponding to the above relative risks are: 9, 900, and 900,000 cases of disease in the population (10 – 1 = 9, etc.). Thus, a large relative risk may apply to very few people, and, conversely, a small relative risk can apply to a very large population. The contrast in the association of smoking with lung cancer and heart disease points up the difference between relative and absolute risks. Compared to those who have never smoked, current smokers have a relative risk for lung cancer of roughly 20 but a relative risk for heart disease of roughly 2.0. However, in spite of the much larger relative risk for lung cancer, smoking is actually responsible for causing more illness through its effect on heart disease than through its effect on lung cancer. This is due to the fact that heart disease is a much more common disease than lung cancer.

The importance of putting information regarding relative risks in perspective by presenting complementary information on the absolute risk and population impact has recently received attention both in medical and public health journals and in the lay press.[44–46] A recent review of the use of relative risks and absolute risks in medical publications concluded that, "Effects presented in relative terms alone have been repeatedly shown to seem more impressive than the same effects presented in absolute terms in studies of physicians, policy makers, and patients. . . . The lack of accessibility of these fundamental data [i.e., absolute risks—G.K] may well lead journal readers (doctors, policy makers, journalists, and patients) to have exaggerated perceptions of the reported effect sizes."[47] As one commentator has remarked, there is an "essential tension" between the perspective of relative risk and that of absolute risk.[48] Each provides crucial information in gauging the importance of a particular cause or risk factor within the context of the overall health of a population.

In addition to its role in identifying specific risk factors with the ultimate goal of preventing specific diseases, epidemiology provides another valuable function—that of monitoring the health of populations and monitoring the prevalence of risk factors. From this perspective, based on existing knowledge, we can project, for example, that the increase in the prevalence of overweight and obesity in the United States and elsewhere is likely to have substantial consequences in terms of the disease burden due to diabetes and other chronic diseases related to overweight. Monitoring of this sort can help inform intervention strategies designed to reduce the prevalence of risk factors in order to reduce the disease burden in the future. Epidemiology has also been able to allay fears concerning many suspected hazards which have arisen over the past thirty years, including—to mention just a few—possible adverse effects of coffee drinking, chocolate, silicone breast implants, oral contraceptives, cell phones, and electromagnetic fields.

Although the tendency of epidemiology to focus on individual risk factors has been extremely successful and has yielded vital new knowledge, the limitations of what has come to be referred to as "risk factor epidemiology" have prompted some epidemiologists to delineate a broader and more encompassing perspective. This alternative view emphasizes the complex "web of causation" and the multidimensionality of the social and material environment. It is, of course, important to document the many health effects of a behavior like smoking, but it is also important to recognize that smoking habits are often correlated with many other behaviors and exposures that affect health. For example, smokers tend to consume more alcohol and coffee, to eat less healthy diets, to engage less in physical activity, to have more sexual partners, and to be less likely to use seatbelts than nonsmokers. In the United States, smoking also has increasingly become a marker for lower socioeconomic status. And socioeconomic status has been shown to have a surprisingly strong effect on mortality, which is independent of risk factors like smoking and obesity.[49]

This broader, more encompassing perspective, which has been referred to as "eco-epidemiology," emphasizes the importance of viewing human health as multidimensional in terms of a whole array of behaviors and exposures in their social and material context, and, at the same time, considering overall health and longevity, instead of just a particular "outcome" of interest. This approach can do much to put the knowledge of specific risk factors and suspected hazards in perspective and can thereby have a beneficial effect on what public health messages are given to the public.

One value of this more comprehensive approach to health and society is that, when it comes to a new potential hazard, emerging evidence is seen against the background of what is known about risk factors for specific diseases and their population impact. We have to keep in mind the essential tension between the information conveyed by the relative risk and that conveyed by the absolute risk. A risk factor with only a modest relative risk can have a large population impact if its prevalence in the population is high and the disease is a common one. This is the rationale for paying attention to subtle and hard to detect environmental exposures, such as exposure to DDT, electromagnetic fields, radon, air pollution, and passive smoking. Because these are difficult to measure exposures, it is all the more imperative to critically evaluate the evidence from all relevant research, to acknowledge the limitations of the existing evidence, and, above all, to put any tentative conclusions in the context of what is known about other causes of the disease. In addition, these more difficult to measure exposures may turn out to affect specific subgroups and to interact with other exposures in ways that will only be revealed by continued research.

But another essential tension, which is equally important, is that conveyed by the contrasting associations displayed in tables 2.1 and 2.2, between risk factors that are firmly established and those that are topics of legitimate study but where the evidence is weak, inconsistent, or inconclusive. Here it is the responsibility of researchers, regulators, editors, and journalists to provide a sober and critical assessment of the limitations of the evidence. Reasonable interim conclusions based on the available evidence may take a number of different forms. For example, the conclusion may be that, in spite of extensive work, studies to date show no consistent or credible evidence of health effects (EMF), while recognizing that this lack of evidence of adverse long-term effects may be due to methodological limitations affecting all studies of a given question (DDT). Or one may conclude that a given risk factor is indeed a proven hazard but that its proven effects are limited to those with exposures above a certain level or among smokers (residential radon). Finally, one may conclude that a given risk factor appears weak and that the existing studies should not be overinterpreted, that the increase in risk associated with exposure may be at the limit of what epidemiologic studies can detect but that nevertheless a weak effect is plausible and that, as a result, exposure should be avoided (secondhand tobacco smoke). These kinds of distinctions can go a long way toward reducing the confusion generated by results that get publicized based on a finding of a "30 percent" increased risk, which is not only small but also uncertain due to the uncertainties of the underlying science. This kind of qualification would help people to focus on firmly established and large population effects—things that will make a difference. The uncertain 30 percent increased risk for lung cancer and heart disease attributed to secondhand tobacco smoke needs to be compared with the firmly established 2,000 percent increase in risk of lung cancer and the 100 percent increased risk of heart disease due to active smoking. Furthermore, a 30 percent increased risk attributed to secondhand smoke is different from a 30 percent increased risk for breast cancer attributed to hormone replacement therapy since the latter association is based on more reliable exposure information.

As part of a more global perspective on health and society, it needs to be realized that in addition to the measures derived from epidemiologic studies—relative risk and absolute risk—there is a very real risk and very real consequences associated with overstating the findings and implied health implications of a suspected health hazard. These entail confusing the public as to what are truly important health risks, leading to needless anxiety and fatalism ("everything causes cancer"; "we are surrounded by environmental hazards") and a diversion of funds away from less sensational (because better established and less novel and anxiety-provoking) causes of disease, entailing a very real cost to society and the public's health. This risk, which has to do with how information from scientific studies is communicated to the public, qualified, and put into perspective, could be called "misinformation risk," and is the topic of the succeeding chapters. By setting findings regarding potential environmental hazards against the background of the solid and far-reaching achievements of epidemiology and its value in monitoring the overall health and exposure status of the population, we will be in a better position to understand how specific environmental threats were overblown and wrenched from their appropriate context.

Box 2.2 Major Take-Home Points

Epidemiology has a rich record of achievements that have the potential to affect the lives of millions of people.

However, epidemiology is an observational science and, for this reason, is subject to the pitfalls of confounding and bias.

Epidemiology deals with populations. Even strong findings that apply to a population as a whole usually cannot be used to accurately predict individual risk.

Findings from epidemiologic studies are often tentative and serve as clues to be pursued by researchers rather than as knowledge that has immediate practical applications.

One needs to distinguish between findings that are well established and that actually matter, in terms of relevance to health, from those that are weak and inconclusive or are simply topics for further study.

Findings that are well established are usually findings pertaining to an exposure that can be measured with reasonable accuracy and that have been confirmed by studies carried out in different populations, as well as by clinical and experimental findings.

Findings regarding a given risk factor for a disease need to be placed in the context of what is known about other risk factors for that disease.

In presenting the results of epidemiologic studies, it is important to provide the absolute risk, a measure of the population impact, as well as the relative risk.

It is important to think about health in a global way, as opposed to focusing only on isolated risk factors for specific diseases. Some exposures are risk factors for one disease but are protective against another disease (estrogen increases the risk of cancers of the breast and endometrium but is protective against osteoporosis. Light/moderate alcohol consumption is a risk factor for breast cancer but appears to be protective against heart disease). Thus, it is the overall pattern of one's lifestyle and environmental exposures that is most important.

3

DOES THE ENVIRONMENT CAUSE BREAST CANCER?

Nothing will ever happen to breast cancer unless it is politicized.

— Fran Kritchek, 1993

Breast cancer is a puzzle of enormous complexity. We don't really have a clue as to what the factors are. What we're trying to do is broaden the question to ask, what is wrong with a society that causes this?

— Susan Love, 1993

Barbara Balaban has been a breast cancer activist since the early 1990s, when, like many other educated women on Long Island, her frustration with the lack of an explanation for why she and others around her had gotten breast cancer pushed her into activism. As a breast cancer survivor and a social worker, Balaban had helped establish the Adelphi Breast Cancer Hotline and Support Program, which provided support groups for breast cancer patients. She was familiar with the literature from the National Cancer Institute and the American Cancer Society describing women who were at increased risk for breast cancer as being over fifty years of age, Jewish, having a poor diet (i.e., high in fat), never having had children, and with a family history of the disease. But the women who came in to the support program did not fit the "high-risk" profile. They were not necessarily older, they had good diets, and they had had children. Most had no family history. It struck Balaban that something else had to be going on.[1]

In the fall of 1990 the New York State Department of Health published a study of breast cancer on Long Island, which had been undertaken in response to community concern, and the results were announced at a press conference. Balaban was one of a handful of people who attended, and she was outraged that, as she put it, "they repeated the same old stuff about 'high-risk groups.'" She recounts how, in the course of the presentation, the health department official displayed a map showing two "high-incidence areas." One was the Great Neck peninsula and the other was the Five Towns area, which consists of Woodmere, Lawrence, Hewitt, Hewitt Harbor, and Inwood. "I was familiar with these areas," Balaban told me.

"They are very heterogeneous, with the very, very wealthy in some areas and the very poor in others. It didn't make any sense. I spoke up, saying this, and it got reported in the papers. This energized people to stand up and say, 'Hey, we've got to get money for research and see that it gets done!'"

More than any other single event, this incident galvanized Balaban to become involved in advocacy for aggressive research into the causes of breast cancer. Within a year or two this came to mean focusing on environmental exposures as causes of the disease. As it was to play out in the course of the 1990s, the high-profile and highly charged issue of breast cancer and the environment reveals a complex web of reciprocal interactions between scientists, activists, politicians, and government agencies that fund research in this area. What studies were undertaken, how they got interpreted and reported in the press, and what the lay public thought was going on were all profoundly influenced by the activists' conviction that something in the environment was causing breast cancer. There were schisms within the scientific community as well as within the activist community concerning what should be studied. Certain facts were distorted, and others were ignored because they did not fit with some groups' agenda. Several isolated results from scientific studies were taken out of context and played up, adding to public concern and stimulating a wave of research studies that in the end provided little support for a role of hypothesized environmental factors in breast cancer. There was much confusion about how to interpret the studies that were carried out and much disappointment about their null results. Were the wrong culprits studied? Were the methods simply inadequate? Or was the whole idea that the environment played a role in breast cancer simply wrongheaded, as many in the scientific community believed? In the end, both scientists and activists learned important lessons from the early, rudimentary studies initiated in response to activist pressure, and in the longer term a much more sophisticated approach to the complex question of what causes breast cancer has emerged. For these reasons, the issue of a possible environmental contribution to breast cancer offers a rich instance of the interplay between science and the wider society.

Breast cancer is the most common cancer among American women, accounting for nearly one third of all cancers. It is also the second leading cause of cancer deaths in women. Over the period from 1940 to 1982 the incidence of breast cancer increased at an average rate of 1.2 percent per year in the United States, but between 1982 and 1986 its incidence rose more steeply, at a rate of 4 percent per year. Most of the increase in the mid-1980s was likely due to increased usage of mammography, but the reasons for the underlying rise over a longer period of time are unknown.[2] Figure 3.1 shows breast cancer incidence among women less than age 50 and women aged 50 and above in the United States over a thirty-year period (based on data from nine of the National Cancer Institute's SEER registries).[3] At the same time,

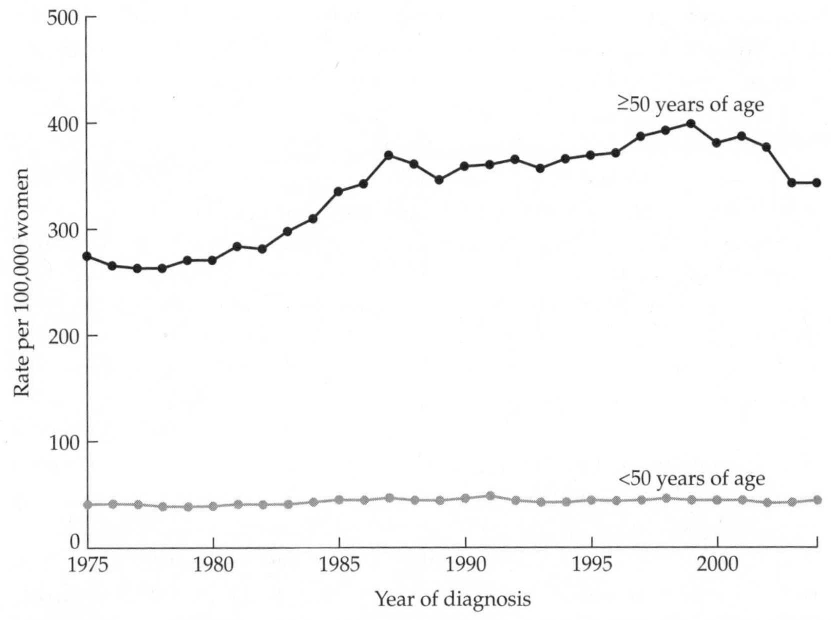

FIGURE 3.1 Trends in breast cancer incidence in women under 50 years of age and in women 50 and older, United States 1975–2004. *Source*: Ravdin et al., 2007.

the incidence of breast cancer was increasing in countries that traditionally had very low rates, and it had become the most common cancer in women worldwide.

Owing in part to the rising incidence rates and the rapid spread of mammography in the late 1980s and early 1990s, breast cancer became the focus of intense concern on the part of survivors of the disease, of women generally, and of politicians and scientists. Yet these developments alone cannot account for the dramatic upsurge of attention to this particular disease at that particular point in time. In the late 1980s the yearly number of new cases of breast cancer was roughly 180,000, and roughly 40,000 women died of the disease. In contrast, heart disease was responsible for about four-and-a-half times as many deaths per year in women, and lung cancer caused 50 percent more deaths in women. Yet neither of these more deadly diseases achieved anything like the visceral hold on women's psyches that breast cancer did.

A number of characteristics of breast cancer help account for the powerful emotions it inspired. First, its impact is magnified because it afflicts women in all stages of life, and the prognosis is poorer in younger women. Furthermore, the disease attacks women in a part of their anatomy that is equated with femininity and beauty, as well as nurturance, and treatment was often radical and disfiguring. Another salient characteristic of the disease is the uncertainty surrounding both its clinical course and its etiology. There is no single overwhelming external cause identified in its development comparable to cigarette smoking, which accounts for roughly 90 percent of lung cancers, or even the handful of risk factors that account for the majority of cases of heart disease. Apparently, the firmer knowledge of the relationship of smoking and lung cancer is less anxiety-provoking than the uncertainty and ignorance surrounding the causes of breast cancer. In addition, breast cancer is a disease of *women*—less than 1 percent of breast cancer cases occur in men—whereas lung cancer and heart disease both have higher rates among men than women. Starting in the 1970s, the long-standing use of the radical mastectomy evoked a strong response from feminists, and subsequently breast cancer became a core issue in the women's movement. These factors help account for the urgency and frustration surrounding the issue of breast cancer that reached their apogee in the late 1980s and early 1990s. Extensive media attention further amplified women's fears of the disease. Although the majority of women who develop the disease survive, a diagnosis of breast cancer was viewed by many as a death sentence. As breast cancer became a high-profile disease, frustration and anger mounted at the lack of knowledge relevant to prevention.

KNOWN RISK FACTORS

As early as 1701, the Italian physician Bernardino Ramazzini had noted that breast cancer was more common among nuns than in married women, and he speculated that something about having children was protective. In the early twentieth century, physicians observed that women whose ovaries were removed at a young age had their risk of breast cancer reduced by half. And over the past 40 years, epidemiologic studies have consistently demonstrated that menstrual and reproductive factors influence a woman's risk. These include an earlier age at onset of menstruation, later age at first full-term pregnancy, later age at menopause, having fewer, or no, children, and not breast-feeding. Each of these risk factors is associated with increased exposure to ovarian hormones, and particularly estrogen. Based on these findings, it became widely accepted that the greater the number of menstrual cycles a woman experiences during her lifetime, with exposure to high levels of estrogen and progesterone, the greater her risk of breast cancer. Being obese modestly increases the risk of breast cancer in post-menopausal women, and this is likely due to the conversion of androgens to estrogen in fat tissue, which becomes the major source of estrogen after menopause. In addition to these factors, a family history of breast cancer in a first-degree relative increases a woman's risk by roughly twofold. A heritable factor underlying family history was identified in the mid-1990s with the discovery of the susceptibility genes, BRCA1 and BRCA2. While mutations in these genes carry a very high risk of breast cancer, because of their rarity, they account for only a few percent of breast cancers.

However, these known risk factors were judged by epidemiologists to account for less than half of breast cancer cases, and individually they showed only modest associations with the disease. Furthermore, these factors were not amenable to modification. What other factors accounted for the large unexplained portion of breast cancer incidence? Diet, and particularly dietary fat, became a major focus of epidemiologic research on the causes of breast cancer in the 1980s. This was due to the strong observed international correlation between intake of dietary fat and breast cancer death rates, displayed in chapter 2. By the mid-1990s the highly publicized notion that dietary fat was an important risk factor had gained widespread acceptance, even though, by that time, large cohort studies had failed to support this relationship. Other potential risk factors that were examined included use of hair dyes, cigarette smoking, alcohol consumption, height and weight, and exposure to X-rays, but these showed either no relationship or only a weak relationship to the disease, or they accounted for only a small proportion of cases. After decades of research on the role of lifestyle factors assessed in adulthood on the risk of breast cancer, little was known that would permit women to reduce their risk of the disease.

At this juncture, in the late 1980s and early 1990s, three widely circulated ideas came together to focus attention on a link between breast cancer and the environment. These were: (1) the perception that there was an "epidemic" of breast cancer in certain areas of the country, such as Long Island, New York; Cape Cod, Massachusetts; and the San Francisco Bay area; (2) the idea that known risk factors could explain less than half of the disease incidence; and (3) the conviction among many survivors and the public generally that some form of environmental pollution must play a role. In actuality, there was no epidemic of breast cancer on Long Island, and there was little evidence to support a role for the environment as a major cause of the disease. Nevertheless, these beliefs became the driving force behind an extraordinary political campaign by breast cancer advocates on Long Island and elsewhere.

Attention to cancer rates was to play a key role in drawing attention to breast cancer as a neglected problem. But, as students of medicine and public health usually learn in their first class in preventive medicine, what constitutes a "high" rate is not always obvious. In the early 1990s, the New York State Department of Health published statistics indicating that breast cancer incidence rates in Nassau and Suffolk counties were slightly higher than those for the state as a whole. For the period 1978 to 1987 the breast cancer incidence rate in Nassau County was 103 cases per 100,000 women, or 16 percent higher than the state average of 89 cases per 100,000. In Suffolk County the rate was 97 cases per 100,000, or 9 percent higher. These rates are similar to rates in other affluent suburban communities elsewhere in the United States, and most epidemiologists believe that they have a straightforward interpretation, which I will come to in a moment. Furthermore, rates fluctuate, and, when data for the 1990s were published, the excess in Nassau and Suffolk had decreased. However, somehow, the figure of a 30 percent excess in breast cancer incidence on Long Island was cited, and once this figure gained currency, it proved virtually impossible to correct.[4] The activists and the media continued to cite this figure in order to maintain a high level of public concern, even though scientists and federal officials who were familiar with the actual rates were alienated by this distortion of the facts. What national statistics showed was that breast cancer rates were generally higher in the northeast of the United States, and Long Island's rate was in line with the region as a whole. There never was an epidemic of breast cancer on Long Island.[5]

To most epidemiologists, the fact that breast cancer rates are slightly higher on Long Island than the average for New York State is not surprising. The population of Long Island is more affluent and more educated than the norm, and these groups tend to have children at a later age and to have fewer children compared to less affluent women. Both of these factors are associated with higher breast cancer incidence. Use of hormone replacement therapy, which modestly increases the risk of breast cancer, and usage of mammography, which leads to a higher rate of detection of breast cancer, are also more common in women in higher income brackets.

The estimate that 50–70 percent of breast cancer could not be explained by known risk factors encouraged the activists in their belief that the "environment" must be responsible for a large part of the unexplained proportion.

The vastness of the scope of possible agents in the water, soil, air, and food that could conceivably make a contribution to breast cancer was equaled by the activists' certainty that there must be an external cause and that that cause must be some form of pervasive pollution. But the existing evidence that some environmental factor or combination of factors played a major role in breast cancer was weak on the face of things. For one thing, breast cancer rates tend to be higher in higher income groups, which in general would be expected to be exposed to *lower* levels of pollution. Also, since the creation of the EPA and the passing of the Clean Air Act and the Clean Water Act in the early 1970s, air and water quality had improved greatly on Long Island and elsewhere. The use of DDT and PCBs had been banned in the United States in the early 1970s, and levels of these compounds in the environment and in human tissues had declined steadily over the following decades. Furthermore, Long Island was not more heavily polluted than many other areas of the country. Finally, animal evidence for the carcinogenicity of organochlorine compounds, including DDT and PCBs—which were to become a major focus of research in the 1990s—was weak, although the evidence for polycyclic aromatic hydrocarbons, or PAH, produced in the burning of fossil fuels, wood, and cigarettes, was somewhat stronger.[6]

Recognition of these facts should not have ruled out study of a relationship of environmental factors to breast cancer, which some scientists felt was justified. But it could have helped to correct fundamental misconceptions of the public and the breast cancer activists and to rein in unrealistic expectations for what conventional epidemiological studies carried out in places like Long Island could achieve. In spite of certain questionable assumptions, the activists had put their finger on a crucial issue, namely, how limited current scientific knowledge is concerning the causes of breast cancer. They were correct in emphasizing the fact that existing knowledge did not allow one to predict with any certainty who would develop breast cancer, much less how to prevent it.

In retrospect, it is striking how disposed the public was to believe that some form of environmental pollution—whether chemicals in the soil and water, radionuclides from nuclear reactors, or magnetic fields from power lines, or something else—must be involved in the development of breast cancer. But, from the beginning, there was a fundamental divergence between the beliefs of the lay public and scientists generally on this question. This can be seen in a Harvard survey from the mid-1990s in which lay women as well as scientists were asked for their ideas on what caused breast cancer. Fifty-six percent of lay women believed that "chemicals" in the environment played a role in the disease, whereas only 5 percent of scientists did.[7] Most breast cancer researchers, including epidemiologists and basic scientists, were simply not receptive to the idea that the environment contributed to the etiology of the disease. Their skepticism stemmed from the fact that, if easier to study exposures like cigarette smoking, alcohol consumption, and hormone replacement therapy showed either no clear-cut relationship or, in the case of alcohol and hormone replacement therapy, only a very modest effect, the very much lower levels of exposure to chemicals in the environment were unlikely to show any relationship. Skepticism on the part of most scientists was to make it very difficult for researchers who felt that there was sufficient justification for carrying out studies to address a possible link to the environment. As one veteran researcher put it, proposals on this topic tended to be "derided" in federal grant review panels.[8] Another reason for the skepticism of breast cancer researchers was that a number of studies had shown that most of the regional variation in breast cancer rates within the United States could be explained by the distribution of known breast cancer risk factors.[9] In the early 1990s, in response to pressure from community activists, the federal Centers for Disease Control had also examined this question and had come to same conclusion.

LONG ISLAND AS LABORATORY

By the early 1990s, Long Islanders' awareness of environmental pollution had been growing for two decades. Long Island is a densely populated coastal formation adjacent to New York City, which underwent rapid population growth and development starting in the 1940s. Over the thirty-year period between 1940 and 1970, the population of Nassau County increased by more than threefold and that of Suffolk County by almost sixfold. In the 1970s, there was widespread concern in the United States about environmental pollution and its possible link to cancer rates. Specific incidents like the Three Mile Island release of radioactive material and the evacuation of Love Canal in the late 1970s seemed to represent only the most extreme examples of a pervasive problem. On Long Island, concern about pollution focused on highly publicized well closings due to elevated levels of chemical pollutants, as well as on landfills, incinerators, and industrial sites. Long Island had been an important agricultural area until the 1950s and 1960s, and pesticides and herbicides, including DDT, chlordane, and dieldrin, had been widely used. Residents of Suffolk County also worried about emissions of radionuclides from the nuclear reactor at Brookhaven National Laboratories. Another source of pollution was the large volume of automobile and airplane traffic. In response to these fears about environmental pollution on Long Island, in 1990 the New York State Department of Health issued the report mentioned earlier, which confirmed the role of known risk factors but failed to demonstrate a clear link to environmental pollution.

Like Barbara Balaban, many breast cancer advocates on Long Island were enraged by what they saw as the failure of state and federal agencies, like the New York State Department of Health and the Centers for Disease Control, and the scientific community to acknowledge that they could not explain what was causing most breast cancers. Many activists had the impression that their neighborhoods contained more women with breast cancer and other cancers than should be occurring normally. In other words, they identified what they thought were clusters, even though it is notoriously difficult to determine whether an apparent cluster really has a common cause, or whether it is merely due to chance, since inevitably, some areas are going to have higher rates of cancer and some lower rates, simply due to chance.[10]

A number of breast cancer survivors undertook ambitious surveys in their communities producing detailed maps indicating homes where an occupant had been diagnosed with cancer. One theory put forward by a prominent activist held that breast cancer tended to occur in homes that were at the end of the water distribution system, suggesting that some pollutant in the water was accumulating in these homes. However, the assessment of geographical clusters requires sophisticated statistical techniques, and these lay efforts failed to produce interpretable results.

Spurred by their conviction that something was going on in their communities and inspired by the recent success of AIDS activists, breast cancer advocates on Long Island formed a number of organizations to raise consciousness in their communities regarding breast cancer and to press for increased government funding for research into its causes. The most influential of these organizations was One-in-Nine (the name referred to a woman's chances of developing breast cancer in her lifetime). Many towns had their own breast cancer organizations, including Babylon, Huntington, West Islip, Brentwood, and Garden City, and these local organizations joined together to form the Long Island Breast Cancer Network. The women in these groups were extremely well-informed about breast cancer, were highly vocal and effective politically, and they had a well-defined goal. They wanted answers to the question of what had caused them, their relatives, and their neighbors to develop the disease and how they could prevent it in their daughters. In 1991 the National Breast Cancer Coalition was formed in order to lobby Congress for increased funding for research. The coalition held several meetings with scientists both in Washington and on Long Island to "brainstorm" about what could be done to address the gaps in knowledge about breast cancer. By bringing together prominent scientists concerned with breast cancer and advocates who had both a commitment to see progress in research and considerable political clout, these meetings provided the impetus for a large infusion of funds into breast cancer research. Barbara Balaban was one of the organizers of these meetings, and she recalls how excited and "energized" both the scientists and the activists were at the prospect of a concerted effort to address the neglected area of environmental contributions to breast cancer.[11] Galvanizing speakers like the breast cancer surgeon and author Susan Love and the epidemiologist Devra Lee Davis helped articulate the activists' sense of purpose and give it legitimacy in the eyes of politicians.[12]

To strengthen their case, activists made a highly effective comparison of the number of breast cancer cases with the number of AIDS cases over a twelve-year period and the amounts of money devoted to research on each disease. The comparison revealed that, in spite of its affecting seven times as many people as AIDS, breast cancer received only a third of the research funds devoted to AIDS.[13]

Once they had an action plan articulated by the National Breast Cancer Coalition, the activists started to lobby politicians to support increased funding for research on breast cancer. In the Senate, Tom Harkin of Iowa and Alphonse D'Amato, who was up for reelection from Long Island, became key supporters of increased funding for breast cancer, as did all of the five representatives from Long Island. Another key player was Phil Schiliro, an aide to Henry Waxman, the chairman of the House subcommittee on health and the environment. Schiliro was planning to run for a congressional seat on Long Island in 1992, and he realized the enormous power of breast cancer as an issue. It was Schiliro who, with input from Devra Lee Davis, drafted the legislation for a study of breast cancer on Long Island, which Waxman and D'Amato then introduced in Congress. After years of media attention to the supposedly high rates of breast cancer on Long Island and to concerns about the environment, the issue was now bipartisan and bicameral. When one congressman asked Barbara Balaban, "Why Long Island?" she responded, "This will be a model for the whole nation."

THE CONGRESSIONAL MANDATE

In response to lobbying by Long Island activists and the support from powerful congressmen, the National Cancer Institute and National Institute of Environmental Health Sciences issued several requests for applications (RFAs), to stimulate researchers to investigate the reasons for the geographic variation in breast cancer rates within the United States.[14] But an even more dramatic response came in the form of Public Law 103-43, which was passed by the U.S. Congress in June 1993. This unusual piece of legislation directed the head of the National Cancer Institute, in collaboration with the head of the National Institute of Environmental Health Sciences, to "conduct a case-control study to assess biological markers of environmental and other potential risk factors contributing to the incidence of breast cancer" in Nassau and Suffolk counties. The law went on to specify "the use of a geographic system to evaluate the current and past exposure of individuals, including direct monitoring and cumulative estimates of exposure to: 1) contaminated drinking water; 2) sources of indoor and ambient air pollution, including emissions from aircraft; 3) electromagnetic fields; 4) pesticides, and other toxic chemicals; 5) hazardous and municipal waste; 6) and such other factors as the director determines to be appropriate."[15] The full text of the law is shown in box 3.1.

Public Law 103-43 represents an uneasy marriage of science and politics. On the face of it, the law brought together the scientific community, the federal government, and the community activists in the common task of shedding light on the causes of breast cancer. In actuality, however, no matter how many community meetings were held on Long Island with the attendance of activists, politicians, government officials, and academic scientists, it is doubtful that the politicians or the activists on the one hand and the scientists on the other were ever really talking about the same study or really understood each other.

Several features of the law merit comment. First, here was the legislative arm of the government not only directing two institutes of the National Institutes of Health (NIH) to carry out scientific research to address a specific problem, but it was going so far as to dictate the type of study to be used—a case-control study. While a lot of science is a response to directed legislation, it is highly unusual for a congressional law to specify the design, the population, and the hypothesis of a study. One drawback of doing this

Box 3.1 Study of Elevated Breast Cancer Rates in Long Island Public Law 103-43, June 10, 1993

Sec. 1911. Potential Environmental and Other Risks Contributing to Incidence of Breast Cancer

(a) REQUIREMENT OF STUDY

(1) IN GENERAL—The Director of the National Cancer Institute (in this section referred to as the "Director"), in collaboration with the Director of the National Institute of Environmental Health Sciences, shall conduct a case-control study to assess biological markers of environmental and other potential risk factors contributing to the incidence of breast cancer in—

(A) the Counties of Nassau and Suffolk, in the State of New York, and

(B) the 2 counties in the northeastern United States that, as identified in the report specified in paragraph (2), had the highest age-adjusted mortality rate of such cancer that reflected not less than 30 deaths during the 5-year period for which findings are made in the report. [Schoharie County, NY, and Tolland County, CT]

(2) RELEVANT REPORT—The report referred to in paragraph (I)(B) is the report of the findings made in the study entitled "Survival, Epidemiology, and End Results," relating to cases of cancer during the years 1983 through 1987.

(b) CERTAIN ELEMENTS OF THE STUDY—Activities of the Director in carrying out the study under subsection (a) shall include the use of a geographic system to evaluate the current and past exposure of individuals, including direct monitoring and cumulative estimates of exposure, to—

(1) contaminated water;

(2) sources of indoor and ambient air pollution, including emissions from aircraft;

(3) electromagnetic fields;

(4) pesticides, and other toxic chemicals;

(5) hazardous and municipal waste; and

(6) such other factors as the director determines to be appropriate.

(c) REPORT—Not later than 30 months after the date of the enactment of this Act, the Director shall complete the study required in subsection (a) and submit to the Committee on Energy and Commerce of the House of Representatives, and to the Committee on Labor and Human Resources of the Senate, a report describing findings made as a result of the study. [An amendment rescinded the 30-month deadline.]

(d) FUNDING—Of the amounts appropriated for fiscal years 1994 and 1995 for the National Institute of Environmental Health Sciences and the National Cancer Institute, the Director of the National Institutes of Health shall make available amounts for carrying out the study required in subsection (a).

is that it limits the investigators' ability to respond with the best possible scientific conception.[16]

The reasons for the choice of a case-control study are clear. This type of study can be carried out in a much shorter time and is also much less expensive than the main alternative, the cohort study design. Interestingly, a number of scientists, including the National Cancer Institute's project director for the Long Island Breast Cancer Study, Dr. Iris Obrams, felt that a case-control study was the wrong approach and favored a cohort study.[17] However, while cohort studies have a number of strengths, they are not superior to case-control studies for reconstructing exposures that occurred in the past.

Second, the scope of the specific topics to be addressed within the law was vast, and the text of the law provided no ranking of the different topics in terms of importance. There was also no discussion of the rationale behind including specific items on the list and no discussion as to what the focus of the efforts should be. The reason for this is that the government was making funds available for research on the broad question of environmental exposures and breast cancer and leaving it up to the scientists applying for funding to make a convincing case in their proposal for what specific questions should be addressed and how. Scientists are used to this "mechanism" of the federal government for stimulating research. But the diffuse "laundry list" of potential environmental hazards undoubtedly reinforced the activists' conviction that many aspects of their environment were a problem. In reality, what the diffuse list reflects is the fact that almost nothing is known about a possible contribution of the environment to breast cancer.

Third, as written, the law required the use of an overarching geographic information system (GIS), a high-powered, computerized system, to supply the framework for estimating individual exposure to the wide array of exposures of interest. However, the use of a GIS in the study of chronic diseases of long latency, such as breast cancer, was, and still is, in its infancy. In addition, there was a huge problem of missing data. (Data on levels of various pollutants in the air, water, soil, and food throughout Long Island over the decades were generally not available.) The GIS, which was sold to the activists by the National Cancer Institute as the ultimate methodology for coming to grips with the health effects of environmental pollution, was in effect a concept without the necessary data to deliver on its promise. Thus, a central requirement of the mandate was totally unrealistic and unachievable within the envisioned period of performance. More than ten years after the enactment of the public law, a first-generation GIS for Long Island was still under development.[18]

It should also be noted that while the language of the law was both broad and ambitious, at the same time, the stated objective was cautiously worded and modest. It was not to "determine" or "identify" "the causes of breast cancer on Long Island" or "the causes of the elevated rates of breast cancer on Long Island," but merely "to assess biological markers of environmental and other potential risk factors contributing to the incidence of breast cancer." This carefully crafted technical language is highly revealing and goes to the heart of the difference in "culture"—and hence understanding—of the community activists and the politicians on the one hand and the scientists on the other.

Although epidemiologists' ultimate objective is to identify new *causes* of disease, in reality, what they work with are entities that can be carefully defined and measured and correlated with the occurrence of the disease. Thus, the scientists were well aware that all they had to work with were "biological *markers* of environmental and other potential risk factors." And they were also well aware of the need for multiple, independent studies of the same question to clarify a hypothesized association. No single study was likely to pinpoint and confirm a particular exposure as a cause of disease. And, given the problems of accurately assessing long-term exposure, even a large number of studies that suffer from the same limitations, can be inconclusive. Finally, the scientists were painfully aware that the ability to investigate a given question hinged on the availability of adequate methodological tools. Most crucially, the ability to investigate specific hypothesized exposures depends on the availability of validated markers of exposure. (Many exposures may merit investigation, but unless one has a marker of long-term exposure, studies are likely to be inconclusive). These considerations, with their implied limitations on what could be studied, were difficult for the community activists to appreciate and to accept. Where the scientists were talking about "markers of environmental and other risk factors," the activists were thinking "causes" and were anticipating dramatic and useable knowledge.

Another aspect of the legislation merits comment. In order to broaden political support for the bill, two other counties—Tolland County, Connecticut and Schoharie County, New York—were included in the final legislation. Both had the highest breast cancer mortality rates in their respective states in the last year for which statistics were available at the time the legislation was being drafted. However, both counties have extremely small populations, and cancer incidence rates in small populations tend to be unstable. In fact, when data became available for the following years, Schoharie County showed the *lowest* breast cancer mortality of all sixty-eight counties in New York State. Nevertheless, the researchers were obligated under the law to conduct a study of breast cancer in that county, even though over a period of several years, they only succeeded in identifying fewer than a dozen cases of the disease—far too few to obtain meaningful results.

Finally, it should be noted Congress did not appropriate any additional funds to carry out the studies on Long Island. The money designated by Public Law 104-43 had to come out of the existing NIH budget. In other words, every dollar spent on the Long Island Breast Cancer Study Project, as it came to be known, was at the expense of other competing NIH-sponsored projects.

Enactment of the 1993 law represented an extraordinary victory for the Long Island breast cancer activists. Through their persistence and political resourcefulness, they had managed to obtain coveted federal funding for a major scientific study on Long Island. But, owing to constraints imposed by the legislation and the large gulf in understanding separating them from the scientists, it was inevitable that the actual study designed and implemented by scientists would fall far short of their expectations. It is hard to convey to the lay public just how limited any individual study of this kind is owing to methodological limitations and the difficulties of obtaining accurate information on exposures at different periods of life. Each study can only address in a highly simplified form a small number of relationships that are isolated from a vast universe of possible relationships. If one measures exposure at the wrong time of life, one may fail to find an important relationship that exists. This is why it is necessary to carry out many different types of studies in different locations and in different populations. But it is also difficult for the lay public to accept just how incremental progress in a difficult area such as effects of environmental pollution on human disease is.

The activists' expectations of what this high-profile, government-sponsored study could achieve were not tempered by any familiarity with the painfully slow process of epidemiologic research on chronic diseases of long latency or with the considerable practical and technical limitations affecting research in this area. They viewed the study as *their* creation since it was owing to their efforts that the law was passed and the funding made available. The fact that the study was to draw participants from their communities encouraged the activists in their belief that it would turn up an explanation for why many of them had contracted breast cancer and that it would enable their daughters to avoid the disease in the future. In retrospect, it is easy to see that the activists' hopes for a dramatic breakthrough in the understanding of the causes of breast cancer were bound to be disappointed.

THE LONG ISLAND STUDY

A number of different scientific projects were funded under the aegis of Public Law 104-43, and these are collectively referred to as the Long Island Breast Cancer Study Project (LIBCSP). Responding to the congressional mandate for a study, in 1994 a group of New York City and Long Island researchers from major area medical centers submitted a detailed application for a large case-control study to the National Cancer Institute and the National Institute of Environmental Health Sciences. This effort was led by Dr. Marilie Gammon of Columbia University's School of Public Health, who had experience carrying out large collaborative studies on breast cancer. After a lengthy peer review process, including a "site visit" at Columbia by a panel of scientists and extensive written critiques of each component of the study, which needed to be responded to, the proposal was funded. The objective was to enroll every newly diagnosed breast cancer case occurring in female residents of Nassau and Suffolk Counties during a one-year period (August 1, 1996, to July 31, 1997). For each case a woman with no history of breast cancer was randomly selected as a "control." In the end, roughly 1,500 cases and 1,550 controls were recruited to the study. Each participating woman provided extensive information on her personal history, collected by means of an in-person interview and self-administered questionnaires; in addition, most women provided a blood and a urine specimen. Because of its size and prominence, the Columbia study itself came to be referred to as the LIBCSP.

The study had two major objectives. The first was to determine whether levels of organochlorine compounds were higher in the blood of breast cancer patients than in the blood of women without breast cancer. These compounds included the pesticides DDT (and its major breakdown product DDE), chlordane, and dieldrin, as well as polychlorinated biphenyls (PCBs). PCBs are compounds found in coolants and lubricants in transformers, capacitors, and other electrical equipment.

The second objective was to determine whether exposure to a class of compounds known as polycyclic aromatic hydrocarbons (or PAH) produced by the burning of fossil fuels and other organic materials was associated with increased risk of breast cancer. This hypothesis was formulated to address the question whether air pollution from heavy automobile traffic on Long Island as well as consumption of grilled foods contributes to incidence of the disease. PAH can bind to DNA forming "adducts," and the resulting DNA damage is thought to play a role in initiating cancer. The plan for the study envisaged measuring these adducts in the blood of women with and without breast cancer.

How strong was the rationale for focusing on organochlorine compounds and PAH as causes of breast cancer? The answer appears to be: not very strong.

The first hypothesis hinged on the finding that some organochlorine compounds can mimic the effects of estrogen, and, as we have seen, estrogen is widely believed to play a critical role in the development of breast cancer, although the precise mechanism is not understood. By the early 1990s, the possibility that "endocrine disruptors" in the environment might be responsible for a wide range of effects in wildlife and humans had become a topic of serious concern.[19] Thus, the study focused on organochlorine compounds because they are widespread in the environment, measurable levels are found in biological fluids in many Americans, and they persist in the body for many years. However, comprehensive assessments of the available animal and human evidence bearing on the carcinogenicity of DDT, PCBs, and other organochlorine compounds that appeared in the mid-1990s concluded that these compounds were unlikely to affect the risk of breast or endometrial cancer "in any but the most unusual situations."[20] These reports noted that evidence from the handful of small case-control studies did not indicate the existence of a consistent risk.[21] In addition, women with relatively heavy exposure from occupational settings did not appear to be at increased risk. Finally, if DDT were an important cause of breast cancer, the marked decline in DDT levels over recent decades should have led to a reduced incidence of breast cancer, which is not the case.[22]

The rationale for the second major hypothesis concerning PAH was that experimental evidence indicated that chemicals produced in the combustion of organic material were mammary carcinogens in animals. Therefore, higher exposure to these compounds present in tobacco smoke, charcoal-broiled foods, and air pollution might show an association with human breast cancer. The researchers argued that the use of PAH-DNA adducts had the virtue of providing both a marker of exposure and a marker of the individual's ability to repair damage caused by these compounds. But the external reviewers of the grant were far from being convinced of the ability of this component to produce informative results. A major problem with the hypothesis that PAH play an important role in the development of human breast cancer is the failure of a large number of studies to show a consistent effect of cigarette smoking on breast cancer, since smoking and eating charbroiled food are much more significant sources of PAH than air pollution. In addition, PAH-DNA adducts only provide an intermediate-term marker of exposure. They may reflect exposure over recent months but not exposure years earlier, when breast cancer was likely to be initiated. It is also noteworthy that the level of PAH-DNA adducts do not show the kind of strong and consistent relationship to lung cancer risk that one sees with smokers' reports of the number of cigarettes they smoked per day.[23]

A number of other hypotheses were included in the Columbia University proposal, but it is fair to say that these two were responsible for its being funded. (Another major hypothesis—that exposure to electromagnetic fields in the home increased the risk of breast cancer—was the focus of a linked study carried out at the State University of New York at Stony Brook. This study will be described in chapter 4.)

In spite of the LIBCSP's relatively weak hypotheses and its methodological limitations, both NCI and the lead investigators indulged in the kind of public relations promotion of the study that could not fail to encourage unrealistic expectations. According to Iris Obrams, at the time NCI's chief of extramural programs in epidemiology and director of the Long Island project, the study represented "an opportunity for groundbreaking epidemiologic research that may serve, ultimately, as a research model for the nation. When the project is finished, we hope to be in a position to give solid preventive advice not only to women on Long Island, but to women all over the country."[24] Commenting on the package of studies to be carried out under the umbrella of the Long Island Breast Cancer

Study Project, a staff writer for the Long Island newspaper *Newsday* wrote: "Many of the studies will be groundbreaking—either because they will be the first, among the largest and most rigorous or because they are taking a different approach. And, taken together, they should provide an in-depth look at the impact of the environment on breast cancer on Long Island, as well as potentially offer insights into the disease for all scientists and doctors."[25] And Marilie Gammon, the lead researcher of the large case-control study, offered the following: "This study is the envy of the nation. Every women's group wants it, and Long Island has it."[26] Although the euphoria was understandable, given the hard work on the part of all parties in obtaining funding for the study, these statements from people in a position to know better betrayed a naïve expectation that this one set of studies with highly focused objectives and limited methods was going to deliver a breakthrough. It was as if, in the flush of launching the study, both NCI and the researchers succumbed to the naïve expectations of the activists.

The scientific project within the LIBCSP, which the reviewers judged to be of greatest interest, was the organochlorine component. And the major reason for this was that Mary Wolff of the Mount Sinai Medical School, a highly respected toxicologist who had done extensive work on organochlorine compounds, was the lead investigator on that project. Together with Paolo Toniolo, a New York University epidemiologist, Wolff had recently published a study purporting to show a dramatic association between DDT/DDE exposure and breast cancer.[27] This study appeared at a crucial juncture—April 1993—in the mounting concern over environmental pollution and breast cancer. Published in the prestigious *Journal of the National Cancer Institute*, it seemed to provide solid evidence in support of the breast cancer activists' contention that the environment on Long Island played a role in their developing breast cancer. In addition, it strengthened the position of epidemiologists making a case for focusing on the environment and led to the conduct of numerous studies of DDT and PCBs by other investigators.

Wolff and Toniolo had used stored blood samples from New York University's Women's Health Study—a cohort study designed to investigate the role of diet and hormones in the development of cancer—to measure DDT, its main metabolite DDE, and PCBs in breast cancer cases and in women without breast cancer. The Women's Health Study had stored blood specimens obtained from 14,290 women enrolled between 1985 and 1991. From this cohort, the investigators selected 58 women who had developed breast cancer in the one to six months following enrollment and 171 matched control women from the cohort without breast cancer. The authors reported that, after adjustment for potential confounding factors, women with the highest blood level of DDE had a statistically significant fourfold increased risk of breast cancer compared to women with the lowest levels. In contrast, no significant association of elevated PCB levels with breast cancer was observed. In the discussion section of the paper, the authors wrote, "Our data suggest that organochlorine residues and, in particular, DDE are strongly associated with breast cancer risk."[28] And their closing sentence drove home the significance of their results: "Given the widespread dissemination of organochlorines in the environment, these findings have immediate and far-reaching implications for public health intervention worldwide." A similar sentence concludes the abstract of the paper, which is all that many people read.

Looked at in hindsight, perhaps the most striking feature of the Wolff and Toniolo paper is the rather substantial gulf between the limited data presented by the authors and their far-reaching claims for their potential significance. Among the study's strengths were its somewhat larger number of cases compared to previous studies and the authors' ability to take into account potential confounding factors. Another apparent strength was its prospective design—the fact that blood samples were collected from apparently healthy women before the diagnosis of breast cancer. However, the selection of cases diagnosed within one to six months of enrollment makes it likely that the cases already had breast cancer when their blood was drawn, raising a question about whether the levels measured in cases were truly reflective of their predisease levels. Furthermore, bloods were obtained in the late 1980s and early 1990s, and, therefore, organochlorine levels might not be indicative of what the levels were two decades or more earlier, before DDT use was banned.

Although the number of cases was larger than in the handful of previous studies, fifty-eight cases is still small when one wants to look for a dose-response relationship and when one has to take into account a variety of confounding or modifying factors.

A look at the key results demonstrates how an influential conclusion can hinge on tenuous data. In fig. 3.2, taken from their paper, the authors presented graphs for both DDE and PCBs, showing how a woman's blood level of each compound affected her risk of breast cancer. Blood levels of both compounds were partitioned into five levels, or "quintiles," and odds ratios for breast cancer were computed for quintiles 2–5 relative to the lowest quintile (Q1). The authors contrasted the results for the two compounds. In describing the PCB results (panel B), they noted that, in spite of the fact that two levels (Q2 and Q3) were significantly elevated, the "flattening out" at higher levels did not suggest a dose-response relationship. In contrast, they reported a significant increasing linear trend with increasing DDE level (panel A). But comparison of the two panels reveals that the two curves are quite similar. In both graphs the middle point (Q3) represents the maximum, with Q4 and Q5 showing a relative decline. In the DDE data, only Q3 has a statistically significant odds ratio, and the same flattening out occurs at the highest exposure levels. In addition, when the data were categorized more finely into ten levels of exposure, the odds ratio for the highest "decile" (90th percentile) was not substantially higher than the odds ratio for the highest quintile of exposure. This actually provides further evidence of a

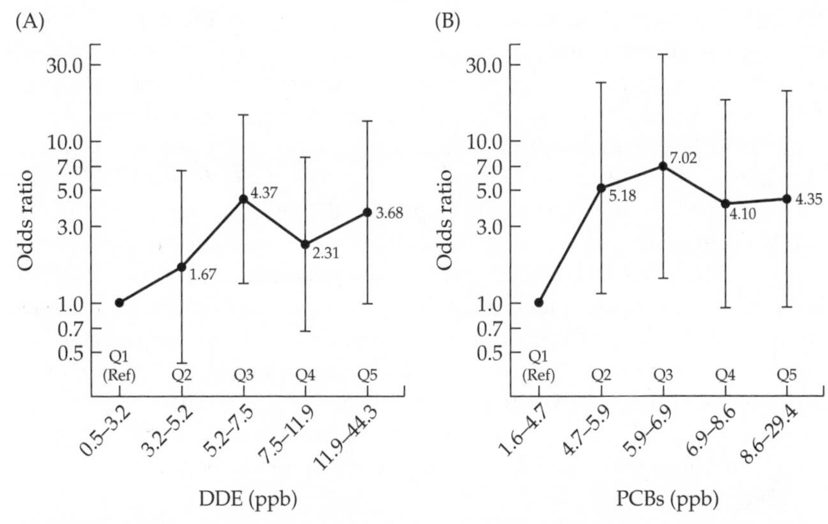

FIGURE 3.2 Odds ratios for breast cancer by quintiles of DDE and PCB concentrations in serum. *Source*: Wolff et al., 1993.

lack of a dose-response relationship. All of this suggests that the results in panel A could merely be a chance finding based on small numbers.

While the authors mentioned the small number of cases and the short interval between blood draw and diagnosis in their discussion, these qualifications were given short shrift compared to the emphasis on the strength of the observed relationship and its implications. Given these aspects of the study, and the problem of using a one-time measurement of DDE in midlife to characterize a woman's exposure in the past, Wolff and Toniolo's claim of a "strong association" and of its important implications for "public health intervention worldwide" was questionable, to say the least.

However, in spite of their study's weaknesses, Wolff and Toniolo's dramatic results galvanized researchers and the federal government to undertake larger studies of organochlorine compounds and breast cancer. The editorial that accompanied the article referred to it as a "wake-up call for further urgent research,"[29] and the National Cancer Institute and the National Institute of Environmental Health Sciences set up special programs to encourage research on DDT and other organochlorine compounds and breast cancer. In the eleven years following publication of the Wolff and Toniolo paper, approximately thirty epidemiologic studies were published on this topic.[30]

Overwhelmingly, the many new and larger studies showed no association of DDT/DDE or PCB exposure and risk of breast cancer. A 2002 review of the accumulated literature by researchers at the American

Cancer Society concluded that: "With rare exceptions, there is consistent evidence from the many methodologically sound studies of no association between levels of persistent organochlorine compounds, notably DDE and PCBs, and breast cancer."[31] Furthermore, a meta-analysis that included 22 studies of DDT exposure and breast cancer reported no elevation in risk of breast cancer (summary odds ratio = 0.97, 95% CI 0.87-1.09), and the authors were unable to explain elevated findings like those of Wolff and Toniolo. Interestingly, the cumulative plot showing the summary estimate with the publication of each new study shows that the summary odds ratio is elevated in the Wolff study but thereafter declines remaining around 1.0, with increasingly narrowing confidence intervals as the weight of more studies is added to the meta-analysis (fig. 3.3). The authors suggested that Wolff and Toniolo's result might just be a chance finding, concluding that, "Overall, these results should be regarded as strong evidence to discard the putative relationship between *p,p'*-DDE and breast cancer risk."[32]

Among the new studies were updated results from the NYU Women's Health Study published in 2000. In the new analysis, based on an average of two-and-a-half years of follow-up (from blood collection to diagnosis) and 148 breast cancer cases and 295 matched controls, Wolff, Toniolo, and

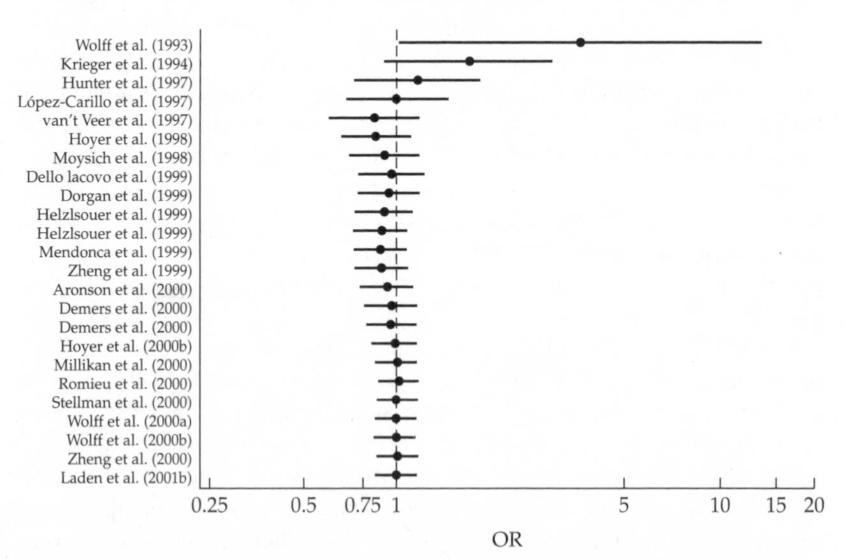

FIGURE 3.3 Cumulative meta-analysis of studies of DDE exposure and breast cancer. *Source:* Lopez-Cervantes et al., 2004. Lines on either side of the black circles give the 95 percent confidence interval. The wide confidence interval of the Wolff et al. study reflects its small sample size. As more studies are added to the meta-analysis, the summary odds ratio approaches 1.0, and the width of the confidence interval becomes narrower.

colleagues wrote: "We found no association of DDE, PCBs, or their half-lives with risk of breast cancer in this cohort study . . . our results do not support a relation between DDE or PCB levels and breast cancer in a prospective cohort of New York City women."[33]

When the long-awaited results of the Long Island study were finally published in 2002, they too showed no association of blood levels of DDE, chlordane, dieldrin, or of the four most common PCB congeners with breast cancer and no indication of a dose-response relationship.[34] Furthermore, there was no suggestion of an effect of exposure in subgroups of women who might be at increased risk: those who had not breastfed, were overweight, were postmenopausal, or were long-term residents of Long Island.

In retrospect, the significance of the 1993 paper by Wolff and Toniolo is that it shows how even highly competent scientists can get carried away by a result and that the scientific community can then react with insufficient skepticism. Not surprisingly breast cancer advocates latched onto this study, since here were two well-known scientists implying that their results provided strong evidence of an environmental exposure on breast cancer. Of course, it is only human to want to come up with positive findings and to have them be meaningful. But for this reason, it is important to be aware that it is only too common for a small, early study of a new question to show an intriguing result. When the initial study is followed by larger and more carefully executed studies, it is often the case that the initial finding is not borne out.[35] This has happened in an early study indicating that coffee drinking was associated with pancreatic cancer,[36] in a study suggesting that consuming beverages containing caffeine could lead to benign breast disease,[37] and in early studies linking fat intake with increased breast cancer risk[38]—to cite but a few instances of this pattern. When the stakes are high—both in terms of the public's desire for knowledge and of researchers' drive to obtain meaningful findings—the danger of getting carried away and committing the cardinal sin of "believing one's hypothesis" is at its greatest.

REACTIONS TO THE LONG ISLAND STUDY

Well before the results of the LIBCSP became known, the breast cancer advocates who had lobbied so aggressively for it had become disillusioned with the project. In part, this was because a considerable portion of the funds from the National Cancer Institute earmarked for research on Long Island had nothing to do with the environment but were to be used to promote mammography and to study medical records. But the main reason for their disaffection was the fact that the case-control study was narrowly focused on only a small number of compounds measured in blood and that these compounds for the most part had been taken out of use decades earlier. There were many other compounds that the activists wanted to see included within the scope of the study, and they hoped that incriminating evidence would lead to statewide bans on additional chemicals. In short, they wanted a much more comprehensive study. However, for the many additional compounds the activists were concerned about, there was simply inadequate information available of their carcinogenicity in animals.[39] In addition, there was no suitable human biological marker that would provide an indicator of long-term exposure. This became the major source of dissatisfaction among the activists. By the time the results were published, they were prepared for the null findings and felt that the LIBCSP had done little to address their concerns.

Publication of the Long Island results in 2002 was greeted with criticism and postmortems from a number of quarters. The Long Island newspaper *Newsday* published a special series of articles delving into the study's history and attempting to explain what went wrong.[40] The *New York Times* published an article by the science reporter Gina Kolata entitled "The Epidemic That Wasn't."[41] One of the figures interviewed by Kolata, Dr. Deborah Winn, the head of extramural epidemiology at the National Cancer Institute, was not exactly supportive of the study that her institute, along with the NIEHS, had spent thirty million dollars on. When asked if the study was based on a false premise—since the breast cancer incidence rates on Long Island were not as high as they had been made out to be—Winn responded, "You're not going to get me to answer that question."[42] Also interviewed by Kolata was Michael Bracken, a professor of epidemiology at Yale University, who was even more critical of the study. According to him, it should never have been undertaken: "It is an example of politicians jumping on the bandwagon and responding to the fears of their local population without really thinking through what is going on in science." Such a study, he said, "is not so much science as a political response."[43] While this is clearly true, it should also be said that the Long Island study was not inferior to similar studies that were carried out elsewhere in the United States.

After almost a decade of carrying out high-profile research on Long Island, Marilie Gammon has a highly nuanced view of how scientific issues can become distorted and politicized when the anxiety of the public is aroused. But she also has keen insights into how, within science itself, there are "fashions" and political pressures. She has an appreciation for the paradoxical nature of the Long Island Breast Cancer Study Project and forthrightly acknowledges that it would never have been funded if it had not been mandated by congressional legislation. On the other hand, even though the study owed its existence to lobbying by activists, Gammon believes strongly that it was justified both on scientific grounds and as a response to the intense level of public concern. As to the scientific grounds, she points to animal experiments that suggested that organochlorine compounds and polycyclic aromatic hydrocarbons (PAH) could play a role in breast cancer. Gammon noted that scientists routinely disagree about how compelling the evidence from specific animal studies is and how it relates to human beings. But she felt that the existing studies were sufficiently suggestive to merit carrying out epidemiologic studies. As she put it, "if you actually look at the data—the laboratory research—there is some hint that something might be going on. . . . Scientifically, I felt that there was enough biological evidence there to make us really take a serious look at what was going on."[44] But she also makes a totally separate argument that, when the public is deeply concerned about a public health question, researchers who, after all, are supported by taxpayers' dollars have some obligation to attempt to resolve the issue. As she put it, "It's hysteria in a lot of ways—you could label it that. But on the other hand, you could say, 'Well, why don't we find out once and for all?' And we can eliminate it if that's really true. So I felt that, even though it was politically-motivated and it wouldn't have happened without politics, I still think it was valuable. . . . Without the politics, it would never have happened. And I don't think that's wrong, I guess is what I'm saying."

Gammon went on to argue that to acknowledge the fact that the study was politically motivated does not mean that its quality wasn't high. On the level of the hypotheses that were proposed and the methods used to test those hypotheses, she feels that she and her collaborators got the science right and that this was acknowledged by her peers.

By deciding to undertake research on a topic that was subject to great public concern and one that was not taken seriously by the majority of mainstream cancer researchers, Gammon had put herself in a thankless position. She made heroic efforts to keep the activists on Long Island informed of the study's progress by giving dozens of talks to community groups and meetings of the Long Island Breast Cancer Network. She tried to correct erroneous beliefs (like the alleged thirty percent excess in breast cancer incidence on Long Island) and to temper their more unrealistic expectations. She tried to convey to them the very real limitations in doing this kind of research. But in spite of her efforts, over time, the activists became increasingly dissatisfied by what this one study could accomplish. Commenting on the gulf between the mindset of the researchers and that of the activists, Gammon said:

> I think that it was harder for them because they expected that if we did the study we would find the right answer. And the truth of the matter is that's not how science works. And in epidemiology it's magnified because it takes a lot longer. You go to the lab and you work for three months and you get an answer. You go into the field and you work for three years, and you still don't have an answer, right? Epidemiology is magnified by whatever percent because it takes so long to do the research. The women wanted a SWAT team. They wanted us to come in and measure

> everything. We didn't know how. We don't know how to measure everything. They wanted Dustin Hoffman in *The Hot Zone* coming in and measuring every window sill. We don't have that. It's a fantasy that we know how to do all that. We don't.[45]

Looking back on the 1993 publication by Wolff and Toniolo, which had provided a major justification for her Long Island proposal, Gammon was refreshingly willing to engage in self-criticism, acknowledging that Mary Wolff and later she herself had gotten carried away:

> I think that our mistake in what we did initially was in assuming that Mary's [risk estimates] could even be that high. Because there's no risk factor for breast cancer that carries that kind of odds ratio. Family history is what—two? So, what were we thinking! That was totally short-sightedness. I remember half-way through the study I turned to Mary and said, 'What were we thinking!' She looked at me and said, 'You're right! That's impossible, you're right!' When you see an odds ratio of four you get carried away.[46]

As we have seen, large epidemiological studies of DDT and PCBs were initiated in the mid-1990s due to the heightened concern about an environmental contribution to breast cancer and, in part, to the provocative results of the article by Wolff and Toniolo. Once a question achieves high visibility and the government puts out requests for applications, researchers flock to address it. However, what needs emphasizing is that when a topic like organochlorines is catapulted into the spotlight, important scientific considerations can be lost sight of. One such issue was the fact that, due to the ban imposed on the manufacture of DDT and PCBs in the United States in the 1970s, levels of these compounds in human tissues and in the environment had shown a dramatic decline over the past several decades. This decline can be tracked in the studies carried out in the United States. A California study used stored blood samples from the mid-1960s; in the 1993 paper, Wolff and Toniolo used stored bloods from the late 1980s to early 1990s; and both the LIBCSP data and another Long Island study used blood samples from the mid-to-late 1990s. And you can see the levels fall from the earlier to the later studies. An important consequence of the declining levels of these compounds over time, which received little attention in most grant proposals, was that the statistical power to detect a dose-response relationship, which depends on having an adequate spread of exposures, was reduced. As Steven Stellman of Columbia University's Mailman School of Public Health put it, "As the range of exposures gets compressed lower and lower, you are getting more noise and less signal."[47]

But a more serious problem confronting all of the studies is one that, again according to Stellman, is glossed over by most researchers involved in this work. This is the assumption that a single measurement of blood or adipose tissue levels of DDT or PCBs at one point in time is sufficient to characterize an individual's lifetime exposure. This assumption was crucial to the rationale for carrying out the epidemiologic studies initiated in the 1990s. At the site visit for the LIBCSP in 1994, Mary Wolff had made the argument that because blood levels of DDT are "in equilibrium" with adipose tissue levels, blood levels can provide an "historical window" on exposure earlier in life at a period relevant to the development of breast cancer. This argument, however, is open to question. Stellman, a physical chemist turned epidemiologist, who has had a long involvement in studying compounds like dioxin and DDT, emphasizes that a one-time measurement of such compounds in midlife can be highly misleading. To illustrate, he frequently borrows an example that was originally used by the Institute of Medicine to criticize the Air Force Health Study of pilots who sprayed dioxin-contaminated Agent Orange in Vietnam.[48] Stellman posits three women who participated in the Long Island study and provided blood samples in the mid-1990s. Suppose that all three women have "low" levels of DDT/DDE in their blood at the time the study is conducted. If the researchers had had one or more measurements from ten, twenty, or thirty years earlier on these same women, when levels of DDT in food and the environment were considerably higher, the results might be quite different. It is altogether possible that in samples taken decades earlier one of the three women might have a "low" value, the second might have an "intermediate" value, and the third might have a "high" value. This is because, owing to individual differences in metabolism and personal history, women may eliminate DDE from their bodies at different rates. Weight loss, breast-feeding, and physical activity may all increase the rate at which these compounds are excreted. Thus, if measurements made later in life and close in time to diagnosis, are used to evaluate the relationship of DDE to breast cancer risk, this may lead to a failure to detect a relationship. Even though many different studies have been conducted in different populations and using different study designs (both case-control and cohort), they may all suffer from the same fatal flaw. This is another way in which the rationale provided by the researchers for conducting the Long Island study—namely that a contemporaneous measurement of DDT/DDE and other organochlorine compounds in blood can provide an accurate indication of exposure decades earlier—was greatly overstated.

Stellman took this issue one step farther. He pointed out that certain researchers have recognized the inadequacy of a single measurement in midlife, and this has led them to address the problem by using the current measurement to extrapolate backward in time using a "pharmacokinetic model." However, not only does the half-life of a compound like DDT (that is, the amount of time it takes the body to excrete half of the body burden) vary between different individuals, but it even varies in the same individual from one time to another. Stellman concluded, "That very observation completely destroys the fundamental assumption of a constant half-life that is the basis of the extrapolation. So, most of the extrapolation that is done now is completely illegitimate. No self-respecting chemist would accept that for one minute."[49]

The implication of these limitations of a one-time measurement of DDT or PBCs in midlife and decades after peak exposure is that the generally null results of these studies do not rule out a possible relationship.

Like Stellman, Mary Wolff also feels in retrospect that the work on organochlorines and breast cancer lacked a strong rationale. She told me that, "frankly you know in the U.S. today all of these exposures are so low, there's just not many heavily exposed people." Furthermore, while she thought that there was evidence suggesting that these compounds might have effects on reproductive health, the same was not true for cancer. Referring to the studies that were done in the 1990s, she said, "Between the susceptibility factors and the decline over time, I'm not sure that these studies are possible. I'm certainly not interested in doing them."[50]

***

Publicity about increasing breast cancer rates in the late 1980s and early 1990s had provided an impetus for societal concern, activism, and research into possible environmental causes of the disease. Activists who focused on a possible role of environmental pollution tended to give little weight to evidence that increased use of mammography and hormone replacement therapy accounted for some proportion of the increase during the 1980s and 1990s. Then, in 2003, national statistics showed an abrupt downturn in breast cancer incidence rates after a six-decade rise (see fig. 3.1). From 2002 to 2003, the incidence of all breast cancers decreased by 7 percent, whereas that of estrogen receptor positive cancers (that is breast cancers that are fueled by estrogen—the most common type in postmenopausal women) decreased by 12 percent.[51] The dramatic reversal in breast cancer rates appeared to mirror the precipitous decrease in the use of hormone replacement therapy that followed the widely publicized results from the Women's Health Initiative in the summer of 2002, indicating that HRT use increased the risk of breast cancer and heart disease. In the six months following publication of the Women's Health Initiative findings, prescriptions for hormone replacement therapy fell by 38 percent. So far, the evidence linking the decrease in HRT use with the decline in breast cancer is purely ecological, and it remains to be seen whether the decrease in incidence persists over a longer period of time and whether it affects mortality. Other possible factors contributing to the downturn include mammography and use of selective estrogen receptor modulators (SERMs) like tamoxifen and raloxifene to prevent breast cancer in women at high risk.[52]

Whatever the full explanation of the downturn in breast cancer incidence, the unexpected change serves to underscore the primary importance of lifestyle factors and preventive strategies in the development of breast cancer. One person who was not surprised by the apparent correlation between the drop in breast cancer incidence and the abrupt decrease in use of HRT was V. Craig Jordan, the vice president and scientific director of the medical science division at the Fox Chase Cancer Center in Philadelphia. For over thirty years, Jordan has played a major role in documenting the effects of estrogen-blocking drugs on breast cancer. As a result of his work and that of others, it is known that tamoxifen, which blocks the effects of estrogen in the breast, reduces the risk of breast cancer in women by about 40 percent.[53] While Jordan points out that removing postmenopausal hormones would not be expected to eradicate estrogen receptor-positive breast cancers, he thinks it is possible that "many subclinical cancer cells may never grow inside a woman's breast if she has no estrogen around to fuel that fire."[54] And, when asked about a possible role of chemicals like DDT or chemicals in plastics that can mimic estrogen in causing breast cancer, his response helps to put this possibility in perspective: "There are a group of compounds like DDT that are byproducts of industry and are in our environment. They can affect cells in the laboratory and can affect the reproduction of animals, but in really huge doses. There is an effect, but does it cause an increase in cancer? I personally don't think there is enough around to do that. A pinch of estrogen in the environment is very small compared to the gallons in a woman's body."[55]

***

In retrospect, the focus on specific environmental pollutants without giving adequate weight to what was known about breast cancer can be seen for what it was—a response to intense public concern based on a number of fundamental misconceptions. Also in retrospect, the Centers for Disease Control were correct in their judgment that the breast cancer rates on Long Island did not justify the carrying out of a "crash" study there to identify the environmental causes of breast cancer. By responding to the alarm on Long Island with just such a crash research program, politicians, individual scientists, and the federal agencies only validated and encouraged the activists' certainty that the environment must be a major culprit. The politicization of breast cancer led initially to the carrying out of studies based on weak hypotheses and inadequate methods, which were greatly oversold by their sponsors. When the results of the studies were published, the activists felt betrayed by the failure to uncover some dramatic and useable knowledge about how to prevent breast cancer.

Much of the difference in perspective between mainstream breast cancer researchers and the activists comes down to how each group defined "environment." To the activists, "the environment" was restricted to chemical and other forms of pollution in the air, soil, water, and food. What concerned them most were pesticide and other chemical residues, products of combustion of fossil fuels, heavy metals, and electromagnetic fields. In contrast, since the beginnings of cancer epidemiology in the 1960s, scientists have defined the environment in the broadest possible terms to include diet, exogenous hormones, and lifestyle behaviors, such as smoking, alcohol consumption, physical activity, body weight, and occupational and other exposures. In this view, the environment includes any agent that can affect the internal milieu, not just "pollution." For scientists, environment is simply contrasted with genetic makeup. To ignore, as the activists did, what was known from decades of research into the relationship of reproductive and hormonal factors to breast cancer, was to ignore an important part of the scientific picture. If the initial studies carried out in response to public pressure had a conceptual failing, it was the assumption that a single measurement of markers of exposure close to the time of diagnosis could shed meaningful light on the risk of breast cancer.

However one judges the initial response to concern about the environment, the cooptation of science by politics was short-lived. Collectively, these studies were able to provide reassurance that organochlorine compounds and PAH—at least as measured in midlife—did not play any substantial role in breast cancer. But their deeper significance is that, once conducted, they made clear the need for more sophisticated approaches to come to grips with the complex biology of breast cancer and possible scenarios whereby environmental exposures might contribute to the risk of disease. This could only be achieved by acknowledging the complexity of breast cancer and of the environment, properly defined.

A decade after the peak of alarm concerning environmental causes of breast cancer, discourse about the causation of this complex disease has become much more sophisticated. One major change is that through organizations like the National Breast Cancer Coalition breast cancer activists have become much more knowledgeable about the science relating to breast cancer and, in an unprecedented development, community representatives participate on an equal basis with scientists in the peer review process for the Army Breast Cancer grants. This means that, for the first time, breast cancer survivors and activists feel that they have a direct voice in the research process devoted to advancing knowledge of how to prevent the disease. On another level, due to the many workshops and consensus meetings that have taken place in the past ten years, there is a greater awareness of the need for a unified view of breast cancer that incorporates what is known from different disciplines, including epidemiology, genetics, endocrinology, toxicology, and molecular biology. Increasingly, researchers are turning their attention to events that occur in the first two decades of life before the breasts are fully developed. And prospective studies are seen as indispensable to document exposures and their effects on early development. In addition, researchers are focusing more on "intermediate endpoints," that is, earlier landmarks preceding the clinical diagnosis of breast cancer, such as increased mammographic density and benign breast disease. Finally, genetic factors affecting susceptibility and the role of gene-environment interactions have become a major focus of research.

Findings regarding environmental exposures need to be integrated into the evolving and highly complex picture of the determinants of breast cancer risk. This means that possible effects of residues from pesticides, products of combustion, heavy metals, or other pollutants, or synthetics, such as bisphenol A, styrene, and other compounds of interest are to be evaluated in the context of their effects on hormone levels, growth factors, receptors, and genes involved in susceptibility to breast cancer. In this new climate, it is less likely that an isolated finding lacking credible grounding in the biology and epidemiology will get the kind of attention from the media and from scientists that was devoted to organochlorines compounds.

Box 3.2 Major Take-Home Points

Much is known about factors that influence the risk of developing breast cancer, but this knowledge cannot be used to accurately predict who will develop the disease and who will not. Furthermore, most known risk factors for breast cancer are not susceptible to prevention.

Within the United States, breast cancer rates are higher in the Northeast and the San Francisco Bay area. This geographic variation in rates is largely explained by sociodemographic and behavioral characteristics. Breast cancer rates tend to be higher in higher income women, who are more likely to have a late age at first birth and fewer children and are more likely to have used hormone replacement therapy. All of these characteristics are associated with increased risk.

In the early 1990s, breast cancer activists became alarmed at what they believed were abnormally high rates of breast cancer in places like Long Island, Cape Cod, and the San Francisco Bay area. Some activists were convinced that some form of environmental pollution was responsible for their having gotten the disease, in spite of a variety of considerations that weighed against this possibility.

A large, federally funded study was undertaken on Long Island to determine whether exposure to organochlorine compounds (including DDT) or combustion products was more common in women who had had breast cancer compared to women who were free of the disease. The results obtained after seven years showed no evidence that any of the hypothesized exposures were related to breast cancer risk. The activists were disappointed at the "null" findings.

The results of the Long Island study and those of other similar studies do not preclude the possibility that some environmental exposures could make a contribution to breast cancer. However, there are reasons to believe that any role of pollution is small relative to known risk factors.

Breast cancer is a complex disease, and current research is attempting to unravel how early events and exposures during and after puberty, as well as genetic susceptibility, can affect risk in addition to known risk factors.

4

ELECTROMAGNETIC FIELDS
The Rise and Fall of a "Pervasive Threat"

Electromagnetic field exposure is so prevalent that the possibility of even a modest increase in disease incidence from this source is worthy of attention.

—David Savitz et al., 1988

In considering the plausibility of biological effects of electromagnetic fields, such an evaluation must depend on quantitative measures of the agent. Only homeopaths will dispute that there must be some level of field strength that cannot possibly affect biology.

—Robert Adair, 1998

With the growth of electrification starting in the late nineteenth century and continuing throughout the twentieth, exposure to weak extremely low frequency electromagnetic fields from man-made sources, unknown until slightly more than a century ago, has become virtually ubiquitous in the United States and other technologically advanced societies. As a result, we are constantly exposed to these fields from the electrical power distribution system, the wiring in our homes and workplaces, household appliances, and industrial and office equipment. It is worth noting that, although we live our daily lives in an intimate relationship with electric power, we are rarely aware of it, except when a major breakdown of the electrical power grid occurs, as happened, most recently, on August 14, 2003, in the eastern United States.

There had been early reports from Soviet scientists in the 1960s of adverse symptoms in workers exposed to very high electric fields, and in the mid-1970s the U.S. government and the electric power industry set up research programs to study the biological effects of electromagnetic fields.[1] However, the idea that exposure to the extremely low frequency electromagnetic fields emanating from power lines and other sources might cause cancer or other serious disease did not become a focus of societal concern until 1979, when a paper entitled "Electrical wiring configurations and childhood cancer" by Nancy Wertheimer and Edward Leeper came out in the *American Journal of Epidemiology*.[2] The reasons for this lack of concern are straightforward. Extremely low frequency electromagnetic radiation is at the low end of the electromagnetic spectrum (fig. 4.1). The energy of electromagnetic radiation is characterized by its wavelength (e.g., in centimeters) or, equivalently, by its frequency (e.g., Hertz, or Hz) because energy (*E*) = Planck's constant X frequency, and frequency X wavelength = speed of light. Thus, the energy of electromagnetic radiation is directly proportional to its frequency and inversely proportional to its wavelength. The lower the frequency, the lower the energy and the less it can affect the cells in our bodies. Ionizing radiation, including gamma radiation, X-rays, and ultraviolet radiation, has extremely high energies that can damage the DNA of living organisms and have other effects on cells. Fields in the microwave range, with very much larger wavelengths (several inches), have enough energy to cause heating in conducting material. Extremely low frequency electromagnetic radiation, with frequencies below 3,000 Hz and with wavelengths of more than 5,000 km, has extremely small energies, which cannot cause heating or ionization. Furthermore, the extremely low frequency electromagnetic fields encountered in everyday life are weaker than the Earth's magnetic field. Finally, the strength of electromagnetic fields decreases rapidly with increasing distance from the source, thus further diminishing the fields to which humans are exposed.

In spite of the inherent implausibility of adverse human health effects stemming from exposure to extremely low frequency electromagnetic fields, the publication of the Wertheimer and Leeper study and subsequent studies throughout the 1980s were seized on by the media and by regulatory bodies, such as the World Health Organization and the U.S. Environmental Protection Agency, as evidence of a potential health hazard. The fact that the fields produced by electric current at 60 Hz (50 Hz outside North America) were both invisible and ubiquitous, that exposure was largely beyond one's control, and that the alleged health consequences were depicted as catastrophic helps to account for the intense fear that came to be associated with this question in the public mind.

The story of how EMF (for the sake of convenience, I will use the terms "electromagnetic fields, "EMF," "extremely low frequency electromagnetic fields," "ELF-EMF," and "power-frequency electromagnetic fields" interchangeably) from power lines and other sources came to be seen as a serious threat to the nation's health sheds light on the way in which preliminary scientific findings can be isolated from their appropriate context, overinterpreted, and given wide currency. In this case, epidemiology by merely following its own internal logic and by generating inconclusive but suggestive results contributed to the creation of an apparent hazard, which was then acted on by society as if it were a real hazard. The social costs resulting from this perceived hazard have been substantial, including not only monetary costs to utilities and the public in the tens of billions of dollars[3]—from the relocating power lines and the abandonment of properties—but widespread and needless anxiety on the part of property owners and parents.

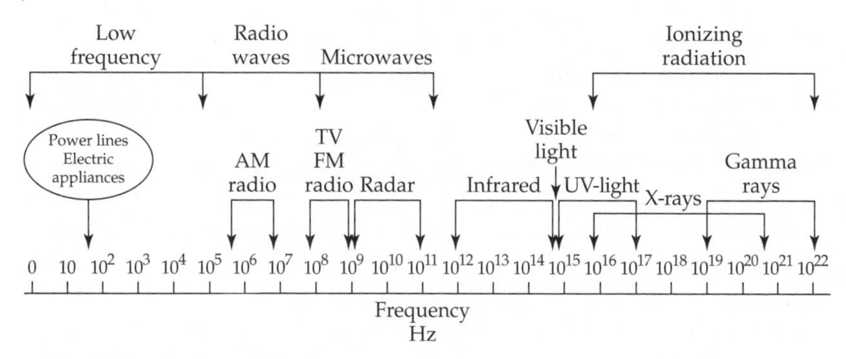

FIGURE 4.1 The electromagnetic spectrum. Adapted from Savitz and Ahlbom, 2006.

The enormous public attention that came to be focused on the question of potential health effects from exposure to power-frequency electromagnetic fields created an long-running drama involving many different players: the electric utilities and their research arm [the Electric Power Research Institute (EPRI)], epidemiologists, physicists, biophysicists, risk assessors and regulators, lawyers, print and television journalists, and the public. Members of these different parties had very different perspectives and stakes in the issue and, to a large extent, could not understand each others' language or assumptions. For this reason, each party saw in the EMF problem what it was disposed to see by its training and narrow objectives. Thus, statements regarding EMF could rarely be taken at face value but, rather, needed to be interpreted in the context of who was making them.

One major axis in the drama was the conflict between concerned property owners and businesses on the one hand and the powerful electric utilities on the other. Owing to its enormous financial resources and political clout, the industry's attempts to discredit early reports of adverse effects from power lines helped to create the impression of a cover-up. Another major axis entails the very different views of the issue held by scientists engaged in research on electromagnetic fields and scientists who were "outsiders" to this field of research and had no particular stake in the issue. This latter group comprised both physicists and a small number of epidemiologists who questioned how much weight should be given to epidemiologic studies of a weak and difficult to measure exposure. Accomplished academic scientists testified on behalf of the utilities—in one case six Nobel laureates (two in medicine) and nine other scientists signed an *Amici Curiae* to the Supreme Court of California to the effect that weak electromagnetic fields were unlikely to harm health—while other scientists, including prominent epidemiologists, testified on behalf of plaintiffs in suits attempting to block the construction of new transmission lines near communities.

Another symptom of how problematic and contentious this issue was is that the U.S. Environmental Protection Agency's 1990 risk assessment report on power-frequency EMF, which determined that EMF was a possible carcinogen, was never released, apparently due to harsh criticism not only from industry but also from independent scientists. Furthermore, shortly after an expert panel concluded in 1992 that power-frequency electromagnetic fields were not a research priority, the U.S. Department of Energy and the National Institute of Environmental Health Sciences launched a special program devoted to basic research and risk assessment in the area of EMF. In this climate, marked by intense public concern, high stakes for a major industry, strong disagreements among scientists, and keen regulatory interest, each new report, regardless of its specific findings or its validity, served to reinforce the perception of a hazard.

Among the scientific disciplines addressing the EMF issue, epidemiology was the lynchpin in the emergence of EMF as a high-profile problem, and it has continued to play a pivotal role. For this reason, in the examination that follows, epidemiology occupies a central position, since the other disciplines and institutions responded—in specific ways and at key junctures and according to their own perspectives—to the epidemiology.

Before considering the epidemiologic studies, a few basics about EMF and units of measurement are in order. Electric and magnetic fields differ in important ways. With respect to possible health effects, most important is the fact that electric fields are easily shielded or weakened by various conducting materials, such as buildings, trees, and even human skin, whereas magnetic fields are not. For this reason, most research into health effects of EMF has focused on magnetic fields.

Exposure to magnetic fields is expressed in terms of *milligauss* or *microtesla*: 1 mG = 0.1 μT. For reference, 1 T is the strength of a typical magnet used in a medical MRI scan. There are 10,000 G in a tesla, and 1,000 mG in 1 G. Thus, the field strength of an MRI magnet is 10^6 mG. In contrast, ambient magnetic field exposures recorded in epidemiologic studies are in the range of hundredths of a milligauss to 10 mG.

THE INITIAL STUDY: WERTHEIMER AND LEEPER

In the mid-1970s, Nancy Wertheimer, a researcher at the University of Colorado in Denver, with a Ph.D. in psychology, became interested in studying the causes of childhood leukemia. Scientists had long suspected that an infectious agent might play a role in the disease, and this led Wertheimer to visit the neighborhoods of children who had developed leukemia in order to study sociodemographic factors and particularly overcrowding. In the course of exploring different neighborhoods, she made the observation that the homes of children who had contracted leukemia appeared to be located closer to electric power lines and substations than the homes of healthy children. With the help of a physicist, Ed Leeper, Wertheimer developed a novel method of estimating residential exposure to electromagnetic fields by characterizing the power lines in proximity to the homes. They distinguished between four exposure levels on the assumption that the greater the distance from the power lines and the thinner the wire, the lower the current flowing through it, and hence the weaker the resulting field. Using this method, Wertheimer and Leeper identified 344 deaths occurring in cases of childhood cancer diagnosed between 1950 and 1973 in the greater Denver area and compared the wiring configuration—or "wire codes"—of the homes occupied by their families with those of the homes of 344 healthy children.

What their data showed was that the homes of children who had developed cancer were between two and three times more likely than the homes of healthy children to have wire configurations suggestive of high-current flow. This was true whether they restricted the analysis to cases and controls who had lived in a single home or in more than one home, and the results appeared to hold independent of potential confounding factors such as urban-suburban differences, socioeconomic class, birth order, sex, and traffic density. The association was seen when different types of cancer were considered individually (leukemia, lymphoma, central nervous system tumors, and "other") (table 4.1). Perhaps most intriguing was the existence of a dose-response relationship when exposure was categorized into four levels ranging from very high to very low.

However, a number of serious problems in the Wertheimer and Leeper study were apparent from the outset and were noted by the authors as well as by other researchers. First, the fact that wire coding was not done in a blind fashion—that is, with the technician performing the mapping of the wire configuration without knowledge of whether a given home was that of a case or a control—could have led to a biased assessment of exposure. Second, as the authors acknowledged, wire codes provide only a "remote proxy" for the actual fields due to power lines, which they were not able to measure. Furthermore, wire codes do not take into account fields due to electrical appliances, ground currents in the home, or other sources outside the home. A third problem was the possibility of selection bias due to differences between cases and controls. If controls are selected in a different manner from cases and represent a different underlying population, this could produce a spurious result. The method of control selection used by Wertheimer and Leeper was so complicated that it was difficult to assess its validity. Finally, the finding that wire codes appeared to be equally associated with different types of cancer in children raises a question since there is no known exposure that elevates the risk of all cancers. As can be seen in table 4.1, when stratified by type of cancer and wire configuration level, the number of cancer cases in each cell were quite small.

In their brief discussion, Wertheimer and Leeper discussed a number of explanations for their findings. They noted that the extremely low frequency electromagnetic fields created by power lines are extremely weak

TABLE 4.1 Wire Configurations and Type of Cancer, Colorado, 1976–1977

Residence	Type of Wiring Configuration	Leukemia	
		Case	Control
Birth address	HCC	52	29
	LCC	84	107
	(% HCC)	(38.2)	(21.3)
Death address	HCC	63	29
	LCC	92	126
	(% HCC)	(40.6)	(18.7)

HCC, high current configuration; LCC, low-current configuration.
Source: Wertheimer and Leeper, 1979.

and that there was no evidence suggesting that they could cause cancer. They appropriately acknowledged that the association observed in their data was modest, on the order of a two- to threefold increase in risk. Wertheimer and Leeper also correctly observed that the lack of specificity with a particular form of childhood cancer militated against a causal relationship, unless EMF acted in such a way as to promote the growth of tumors generally. Yet, in spite of the implausibility of their results and the lack of any supporting evidence, they offered an exceedingly vague rationale for how ELF-EMF might affect physiology:

> AC magnetic fields might affect the development of cancer indirectly, through some effect on physiologic processes. It is conceivable, for instance, that contact-inhibition of cellular growth, or the basic immune reaction of recognizing 'self' from 'not self,' involves electrical potentials occurring at cell surfaces. Against an electromagnetic background different from that provided during evolution, any such cell mechanism might be altered.[4]

Since virtually nothing is known about what causes leukemia or other cancers in children, the possibility that neighborhood power lines and substations played a role was of potentially enormous public health as well as medical significance. The fact that the paper was published in the respected *American Journal of Epidemiology* lent it credibility and caught the attention of epidemiologists interested in the causes of cancer and in preventable hazards in the environment. But many questions are studied by scientists that never gain any notoriety with the public. The potential threat from electric power struck a nerve in a society with a greatly heightened consciousness of pollution from coal-burning power plants and fears about the safety of nuclear power. In this atmosphere, the notion that electrical

Lymphoma		Nervous System Tumors		Other	
Case	Control	Case	Control	Case	Control
10	5	22	12	17	9
21	26	35	45	31	39
(32.3)	(16.1)	(38.6)	(21.1)	(35.4)	(18.7)
18	11	30	17	18	17
26	33	36	49	45	46
(40.9)	(25.0)	(45.5)	(25.8)	(28.6)	(27.0)

current from power lines extending from these plants and reaching into every neighborhood could pose a threat made a kind of intuitive sense. It is noteworthy that when scientists were interviewed in connection with news reports of new findings concerning the health effects of EMF, the producers invariably used an image of ominous-looking high-voltage transmission lines as a backdrop.

Given the serious methodological questions about the Wertheimer and Leeper paper, it is surprising that it received relatively little critical comment in letters to the *American Journal of Epidemiology*. Without much critical discussion, this single paper provided the stimulus for all subsequent epidemiologic studies of EMF and cancer, as well as regulatory interest in EMF. Over the ensuing twenty-five years, a large number of studies were published addressing the issue in a variety of ways, using different types of data, and looking at different disease endpoints. In the early phase of epidemiologic research, the focus was on "hematopoetic cancers" (mainly leukemia but also lymphoma) and central nervous system tumors because Wertheimer and Leeper had found excesses for these cancers in children. These diseases, in adults as well as in children, continued to be a focus of research, while the field widened to include other diseases. Early responses to Wertheimer and Leeper's report included analyses of occupational data, some of which suggested that workers with presumed high exposure to EMF had elevated rates of leukemia and brain tumors. (It should be noted, however, that the predominant type of leukemia occurring in adults is different from that in children.) Then, in the early 1990s, three studies of electrical workers reported an increased risk for male breast cancer, and a modest increase in risk of breast cancer was later noted in a study of women in electrical operations. Since breast cancer was already a focus of intense public concern and epidemiologic research, several large studies of female breast cancer were initiated in the

1990s to address EMF. At the same time, researchers extended their investigations to diseases other than cancer, specifically neurological diseases [including Alzheimer's disease, Parkinson's disease, ALS (amyotrophic lateral sclerosis)], depression, and cardiovascular disease. Other outcomes such as miscarriage were also studied. In this way, epidemiologic research proliferated to include an array of chronic diseases as well as other conditions potentially caused by exposure to EMF. Thus, what began as a single question led to a series of linked but independent questions. There was a continual extension of the object of study to new entities. At the same time, as research progressed, there was a concomitant improvement in methodology and measurement of exposure. In this way, once the question of a hazard from power-frequency EMF became a focus of interest for epidemiologists, it took on a life of its own.

SUBSEQUENT STUDIES IN THE 1980s

In the years immediately following the Wertheimer and Leeper paper, a number of reports on the possible association of EMF with cancer were published in medical and epidemiologic journals. The most important of these was a study from Stockholm, Sweden, that examined the occurrence of leukemia in children and adults in relation to the distance of the residence from high-voltage transmission lines, as well as one-time measurements of EMF at the front door of residences.[5] (In Europe, unlike in the United States, most secondary distribution lines are buried underground.) This study reported a strong positive association with brain cancer in children with exposure to very high measured fields and with historically reconstructed fields but not with leukemia. Several other residential studies showed no effect.

In addition, a number of letters to journals presented information on the mortality and incidence experience of workers with presumed high exposure to EMF, such as telephone linemen.[6] These early occupational studies were even more problematic than the residential studies. First, most of them were based on comparing "proportionate mortality" (that is, the proportion of deaths due to leukemia, or some other cause, as a proportion of all deaths) of workers in jobs presumed to entail elevated EMF exposure with that of the general population or of workers in jobs judged to have no or low exposure. This type of study is difficult to interpret because higher proportionate mortality does not necessarily imply higher mortality rates. Second, all occupational studies lacked direct measurements of fields and, therefore, had to rely on job titles to infer exposure. However, actual exposure could vary greatly among people with the same job title, and tasks as well as exposures are likely to have changed over time. Third, these studies were not able to take into account other concomitant occupational or lifestyle exposures that might affect cancer risk. In spite of the inconsistencies between studies and the many methodological limitations, these studies were taken in the aggregate as providing some support for an association and, therefore, justifying further epidemiologic research.

One consequence of the Wertheimer and Leeper paper was to attract the interest of David Savitz, who in the early 1980s was a young assistant professor of epidemiology at the University of Colorado in the same department as Wertheimer and was looking to develop his own research program. In contrast to Wertheimer and Leeper, neither of whom was an epidemiologist, Savitz had received rigorous training in epidemiology and, for this reason, had the credibility to carry out a study in Denver designed to confirm or refute his colleagues' controversial findings. In the early 1980s, Savitz obtained a grant for a second study of childhood cancer in Denver from the New York State Power Lines Project, which had been set up with funding from the electric power industry. In describing that period, Savitz told me that, to most scientists at the time, the idea that EMF from power lines could cause cancer seemed "flaky," and he himself was ambivalent about studying such a fringe question.[7] Yet little was known on the topic, and, given the nearly universal exposure to extremely low-frequency electromagnetic fields, even if there was only a low probability of health effects, he felt it was justified to carry out careful epidemiologic studies. Such studies, he argued, would provide valuable information needed for decision making by the electric power industry as well as regulatory agencies and the public health community.

Savitz's study, which did not appear until 1988, represented a clear improvement over the original Denver study.[8] He and his colleagues enrolled all 356 residents of the Denver metropolitan area under 15 years of age who had been diagnosed with any form of cancer between 1976 and 1983. Controls were recruited via random digit dialing, a standard technique used in this type of study to obtain a representative sample of the healthy population. Exposure assessment entailed both wire coding (performed by technicians who were unaware of the case-control status of the residence) and spot measurements of electric and magnetic fields in a number of rooms and near the front door. The researchers included all cases of childhood cancer, as opposed to deaths in the Wertheimer and Leeper study.

As reported by the authors, measured magnetic fields under low power conditions showed "a modest association" with cancer incidence. Contrasting homes with fields of 2 mG or higher with homes having fields of less than 2 mG resulted in an apparent 40 percent excess for total cancers, a doubling of leukemias and lymphomas, and a threefold increase in soft tissue sarcomas. No association of measured fields with total cancers was seen under high power use conditions. Contrasting the two highest wire code categories with lower categories yielded a 50 percent excess for all cancers combined.

In spite of its superiority to the original Wertheimer and Leeper study, Savitz recognized that his study had a number of serious limitations. These included poor compliance with the measurement component of the study

(64 percent of cases and 20 percent of controls lacked measurement data), resulting in limited data on specific types of cancer, differential mobility between cases and controls (controls were more residentially stable than cases), and the fact that the measurements taken were only "imperfect surrogates for long-term magnetic field exposure history." Perhaps most importantly, the elevations in risk were not statistically significant, and there was no indication of a dose-response relationship with increasing level of exposure to measured fields (table 4.2). The elevations in risk noted for individual cancers when high measured exposure was contrasted with lower exposure (2 mG or higher versus less than 2 mG), were not even close

TABLE 4.2 Cancer Risk in Relation to Measured Electric and Magnetic Fields, Denver, Colorado Standard Metropolitan Statistical Area

Exposure level	No. of Cancer Cases ($n = 356$)	No. of Controls ($n = 278$)	Odds Ratio	95% Confidence Interval
Magnetic fields (mG): low power use conditions				
0–<0.65	75	134	1.00	
0.65–<1.0	20	28	1.28	0.67–2.42
1.0–<2.5	23	33	1.25	0.68–2.28
2.5+	10	12	1.49	0.62–3.60
Missing	228	71		
Mantel chi for trend = 1.06, $p = 0.14$				
Magnetic fields (mG): high power use conditions				
0–<0.65	61	99	1.00	
0.65–<1.0	23	33	1.13	0.61–2.11
1.0–<2.5	32	54	0.96	0.56–1.65
2.5+	13	18	1.17	0.54–2.57
Missing	227	74		
Mantel chi for trend = 0.19, $p = 0.43$				
Electric fields (V/m): high power use conditions				
0–<6.0	34	54	1.00	
6.0–<9.0	37	67	0.88	0.49–1.58
9.0–<14.0	41	53	1.23	0.68–2.22
14.0+	17	30	0.90	0.43–1.88
Missing	227	74		
Mantel chi for trend = 0.18, $p = 0.43$				

mG, milligauss. *Source*: Savitz et al., 1988.

to being statistically significant, and the estimates were extremely "imprecise" due to the exceedingly small numbers cases and controls in the high-exposure category. Finally, no trend was seen with longer duration of occupancy of the home in which measurements and wire codes were made.

In discussing his results, Savitz carefully described the limitations and weaknesses of his and similar studies. First, consistency was lacking in the detailed findings of the five residential studies of EMF and childhood cancer. Second, the magnitude of the associations found by Wertheimer and Leeper was larger than those observed in his study. Also, Savitz pointed out that "it would be erroneous to interpret the literature as a series of replicated positive results." In other words, the individual studies were different enough in their methods, populations, and results to make it ill advised to interpret them as providing unambiguous confirmation of an effect. In spite of these careful distinctions and caveats, Savitz's study was interpreted by many epidemiologists and lay persons as confirming the findings of the original study, which had raised the issue of power lines and childhood cancer, or at least providing enough suggestion of a possible association to justify carrying out further studies to resolve the "uncertainties."

DIVERGENT ASSESSMENTS OF RISK

By the late 1980s and early 1990s a sufficient number of studies (approximately fifty by 1990) had appeared to enable epidemiologists to review the whole body of evidence. Such critical reviews provide an opportunity to examine the consistency of results from different studies as well as the strengths and weaknesses of individual studies. For this reason, they can be influential in directing further research as well as influencing the regulatory process. For our purposes, they also allow one to gauge the range of interpretations of the EMF issue among epidemiologists roughly ten years after the appearance of the Wertheimer and Leeper paper.

A lengthy review article by Savitz and colleagues published in 1989 critically reviewed the early studies and proposed methods for improving the quality of future research.[9] Savitz's role all along had been to press for increased methodological rigor. He reasoned that if there was a causal link between EMF and disease, as the methods improved, more compelling evidence of a risk from EMF should emerge. First and foremost, Savitz and colleagues argued that the assessment of EMF exposure needed to be strengthened. Use of wire configurations and spot measurements were only crude surrogates for what one wanted to know, that is, long-term exposure to EMF from all sources. Exposure assessment could be improved by making use of personal dosimetry (that is, using personal monitors that record EMF exposures over a period of a day or more) to attempt to validate the surrogate markers (including stationary measurements and wire codes). This could be done in both residential and occupational studies. A second area was the assessment of confounding by means of analyzing data on other factors, which are correlated with wire codes or field measurements and which might themselves be risk factors for cancer. Traffic density, an indicator of air pollution, was one such potential confounding factor. Third, selection bias was a major concern since controls tended to be harder to enroll than cases, and some factor related to participation might be influencing the association of EMF with disease. Finally, Savitz and colleagues emphasized the importance of evaluating the temporal dimension of exposure and testing different assumptions since the relevant timing of exposure for leukemia and other cancers is not known. Since EMF is not a cancer initiator and is not mutagenic, it was considered more likely to act as a promoter or growth stimulator. This could mean either that duration of exposure over a number of years might be a key variable or, alternatively, that exposure shortly before detection of clinical disease might be most relevant.

The program Savitz outlined, and to which he contributed abundantly, made an immense contribution to the strengthening of epidemiologic studies of EMF that took place during the 1990s. But there was an inherent conflict that made it difficult, if not impossible, to effectively advocate further study of the problem and at the same time give a thoroughly tough-minded assessment of the existing evidence. If Savitz was too critical and dismissive of the existing evidence, he might undermine the argument for the need for further research. Thus, it is interesting to note that his interpretation of existing studies, while accurate, is couched in language that betrayed this tension. He details their many methodological weaknesses and notes the inconsistencies among them. And yet, somehow these studies seem to add up to something more, although this "something more" is quite vague:

> Thus, the literature on residential exposures provides some evidence of an effect of electromagnetic fields on childhood cancer. The findings are inconsistent among even the positive studies regarding the magnitude of effect of the specific cancers that might be most affected, but there is some consistency in the general direction of the findings. Adult cancers have been less consistently associated with these exposures.[10]

In the conclusion to the review, Savitz and colleagues refer to "several suggestions that prolonged magnetic field exposure . . . may increase the risk of leukemia and brain cancer." However, they continue, "these studies fall short of providing conclusive indications for a causal association." This is actually quite an understatement. In fact, there is a huge gap between the "suggestions" from the early studies and a strong case for causality. But Savitz, like many epidemiologists, needed to minimize this gap in order to obtain the funding necessary to do the work that he thought was important and justified.

A similar ambivalence is evident in Savitz's references to experimental studies of EMF. He claimed that, "There is now growing evidence of numerous biologic effects at relevant frequencies," even though he acknowledged that none of these effects offered direct evidence of carcinogenicity. Furthermore, it is now known that most of these results have proved to be not reproducible.

In this moot situation, with very weak epidemiological data "suggesting" a hazard from EMF, public perceptions and the potential public health implications took on a decisive role. Here is the conclusion of Savitz's review:

> The potential public health impact of this environmental agent argues strongly for continuing and expanding efforts to confirm or refute the reported associations with cancers in adults and children. Policymakers who address power line right-of-ways, home buyers, and even potential users of electrical appliances face decisions which implicitly include judgements about the likelihood of adverse effects. The present state of uncertainty makes it impossible for society or individuals to make these decisions in a scientifically informed manner.[11]

In contrast to Savitz's review, an article published two years later, in 1991, by Poole and Trichopoulos provides a very different assessment of the question of EMF and cancer and how it should be interpreted.[12] Poole and Trichopoulos gave an even more critical reading of the existing studies than did Savitz, including Savitz's Denver study. Because of its importance and prominence, they devoted nearly two pages to a detailed discussion of that study. A major concern was that differences in socioeconomic status or residential mobility between cases and controls introduced by biased selection of controls could have affected the results in unpredictable ways. Considering the available studies, Poole and Trichopoulos concluded that, "there is little, if any, such coherence with respect to childhood cancers, even though many researchers seem to agree that the strongest evidence concerning hypothetical causation by ELF-EMF comes from the childhood cancer studies." They then considered the possibility that the risk estimates might be underestimates due to the problem of misclassification of exposure (resulting from the inability to accurately measure fields). Poole and Trichopoulos showed that this could in fact be the case. However, they went on to argue that if the relative risks associated with "high wire code" exposure were on the order of 5 or 10 (rather than 1.5 to 3), given the fourfold increase in electric power consumption in the United States between 1950 and 1987, we would be witnessing an epidemic of childhood cancers. In fact, they noted, the incidence of childhood cancers has increased little, if at all, over this time period.

In closing, Poole and Trichopoulos discussed the plausibility of an EMF-cancer link. They referred to the two main types of evidence relevant to assessing the question: that from epidemiological studies on the one hand and from experimental studies in the laboratory on the other. These two types of evidence, they argued, are complementary. The stronger the evidence of an association (i.e., substantial relative risks) in well-designed epidemiologic studies, the less a causal interpretation hinges on a credible biological mechanism. Conversely, if there is strong experimental or other biological evidence of a mechanism, a weak or inconsistent association can be given a causal interpretation. However, neither of these conditions obtained in the case of EMF and cancer. In their judgment, "the notion that ELF-EMF can cause cancer is not impossible or absurd but it is hardly plausible." In conclusion, Poole and Trichopoulos pointed out that research on ELF-EMF had been going on for only slightly more than a decade. This was less time, they observed, than was required for the link between cigarette smoking and lung cancer—a much stronger association—to achieve acceptance. Their final paragraph is worth quoting in full:

> The hypothetical carcinogenicity of ELF-EMF represents an intriguing scientific problem and a potentially important public health issue but, at this stage, nothing more. It clearly is not possible to exonerate ELF-EMF. In order to do so, very large and valid studies showing very little or no association between ELF-EMF and cancer would be needed. This condition is not presently fulfilled. On the other hand, the empirical evidence linking ELF-EMF to cancer is weak and inconsistent. Causal interpretations are not supported by the available biologic data. This is an area in which more and better research, and an atmosphere conducive to dispassionate inquiry, are clearly needed.[13]

What Poole and Trichopoulos managed to convey is an unflinching view of the available evidence and a refusal to go beyond it. At no point did they relax their critical faculties, whether they were considering the evidence from epidemiologic studies, from laboratory experiments, or from changes in electric power distribution and in childhood cancer rates over time. They refused to play up the threat to public health angle because the available evidence was "weak and inconsistent." And they made clear that they deplored the misappropriation of early epidemiologic findings by a sensationalizing media and by regulatory zealots. In accounting for Poole and Trichopoulos's ability to render an impartial and intellectually rigorous judgment on this question, it is pertinent to note that neither of them has been engaged in primary research on the effects of EMF. Thus, their research support is not dependent on their taking a particular position on this issue.

Trichopoulos, who is a professor of epidemiology at Harvard, was also a member of an expert panel that reviewed the existing evidence on the carcinogenic potential of ELF-EMF for Oak Ridge Associated Universities in 1992. The panel reached the conclusion that further research on EMF was not a priority. In a letter to *Science* magazine in 1993[14] the authors of the Oak Ridge report subjected three Swedish studies to scrutiny, pointing out the many inconsistencies in their results, as Poole and Trichopoulos had done for the U.S. studies. In the final paragraph of the letter, the authors attempted to put the whole issue in perspective:

> We have never stated that a causal association between EMF and cancer is impossible or inconceivable; we have indicated that the evidence for such an association is empirically weak and biologically implausible. We have not proposed that research concerning the health effects of EMF be discontinued; in fact, we have indicated areas of some scientific interest that warrant consideration for future research. However, given the decreasing resources available for basic health and science research, we believe that in a broader perspective there are currently more serious health needs that should be given higher priority.[15]

The kind of unrelentingly critical and logical assessment of the evidence presented in these two reports is in stark contrast to the message that was conveyed by epidemiologists involved in primary research on EMF and, even more so, to the message conveyed by the media and regulatory agencies.

ENTER PAUL BRODEUR

In 1989 the investigative reporter and staff writer for *The New Yorker* Paul Brodeur published a series of three articles in that magazine claiming that "radiation" from power lines, electric blankets, and video display terminals posed a hazard to the nation's health and that this danger had been systematically suppressed by the government and industry. Later that year the articles appeared as a book entitled *Currents of Death: Power Lines, Computer Terminals, and the Attempt to Cover Up Their Threat to Your Health.*[16] Brodeur's chatty, fact-filled account of research into these hazards made gripping reading. Through the articles, the book, and his appearances on prime time television, his exposé reached a wide audience and had an enormous impact.

What made Brodeur's account so compelling was the unequal contest he dramatized between a small number of researchers who were convinced that their data demonstrated harmful effects and the monolithic "establishment" consisting of the military, the government, industry, and most of the scientific community committed to denying the existence of a hazard. Brodeur's protagonists were a handful of maverick scientists and medical researchers who were portrayed as selfless and courageous, whereas their opponents in the establishment were depicted as cynical and corrupt. It is revealing that, in introducing his protagonists, Brodeur humanized them by describing their physical appearance and their accomplishments, but when he introduced his villains, their physical attributes and their substantial accomplishments did not merit mentioning. All the reader was told was that a given scientist had testified for industry or authored a report that was allegedly favorable to the government or industry. In this way, the reader always knew what the author wanted him to think of a given scientific figure and how to judge his position. Brodeur needed this device in order to fashion a fast-paced narrative in which every building block fitted in with his thesis. While he was not wrong to be skeptical of the military-industrial complex, he seemed unaware that this is not the only methodological tool needed to decode reality. He saw conspiracy everywhere and saw reality as monolithic. The National Academy of Sciences was as suspect as the Pentagon. Furthermore, the only vested interests he acknowledged were those of big industry and government. He showed no awareness that there are subtler forms of vested interest, including that of scientists—and journalists—in selecting results that support their position.

Brodeur's predisposition to credit evidence of a hazard led him to elevate studies that had not been confirmed and to ignore or impugn the work of outstanding scientists. The resulting tissue of misrepresentations, biased reporting, and outright errors was so extensive that, in 1991, the Committee on Man and Radiation of the Institute of Electrical and Electronics Engineers (IEEE) published a 36-page critique of the book entitled *"Currents of Death Rectified"* in which scientists whose work had been distorted by Brodeur addressed specific points to set the record straight.[17] Brodeur's style was designed to persuade the reader of his ability, as a rigorous investigative journalist, to penetrate to the truth behind biased testimony and obfuscatory government and industry reports. However, as the IEEE report documents in detail, throughout his account he conflated distinct phenomena and failed to make crucial distinctions. For example, he discussed the radiation from VDTs as if it were the same as that from power lines and electric blankets, when in fact they are quite different. At key points, he failed to make a fundamental distinction between electric and magnetic fields. And he confused pulsed radiation generated by radar systems with the phenomenon of amplitude modulation. For all *The New Yorker's* vaunted reputation for checking the facts in its articles, no one detected that Brodeur's whole narrative was vitiated by his willingness to select those findings that supported his position.

While Brodeur may be less than trustworthy as a science writer, ironically he probably did more than any other individual to galvanize research into the effects of EMF. His writings and the general news media, along with the more balanced newsletter *Microwave News,* helped create a level of urgency in the general public and the government to assess the alleged threat from power lines and other sources of EMF. As mentioned above, in spite of the recommendation of the Oak Ridge expert panel that research on EMF was not a priority, in 1992 the Congress mandated the creation of the EMF-RAPID program funded by the Department of Energy and administered by the National Institute of Environmental Health Sciences to promote basic research into the health effects of ELF-EMF. This program focused on laboratory and experimental studies and on carrying out a risk assessment of EMF. In addition, substantial epidemiologic research was funded by the National Cancer Institute, the National Institute of Environmental Health Sciences, and the Electric Power Research Institute (EPRI), which was supported by the electric power industry. The existence of a high level of public concern worked in favor of scientists involved in studying the effects of EMF since it permitted them to remain impartial and sober, merely pointing out that even a small excess risk applied to the large exposed general population would constitute a significant pubic health problem. Epidemiologists could do what they are professionally disposed to do, focus on conducting further research and attempting to synthesize the available data and explain apparent inconsistencies and anomalies. Brodeur and others had created an avid constituency for their work.

ISSUES PREOCCUPYING THE EPIDEMIOLOGISTS

Once the question of the health effects of EMF was raised by Wertheimer and Leeper and subsequent studies in the 1980s, epidemiologists devoted themselves to carrying out better studies of this question. Research focused on the technical difficulties of studying an exposure as complex and variable as ambient EMF, and epidemiologists focused on a small number of key issues.

First and foremost was the task of assessing exposure to electromagnetic fields. As Wertheimer and Leeper had pointed out, wire codes were only a poor surrogate for the current flowing through power lines, and their use assumed a constant level of current, when in fact the actual level was apt to vary over time. In addition, use of wire codes did not take into account other sources of EMF, like household appliances, ground currents, and exposures outside the home. What epidemiologists wanted to know was how informative were wire codes and short-term measurements as markers of long-term exposure?

Other questions preoccupying epidemiologists studying the health effects of EMF included confounding (the possibility that some other factor that was correlated with exposure to EMF was responsible for the observed association between EMF exposure and disease), selection bias (which could occur in case-control studies if the controls are not drawn from the same underlying population as the cases), the timing of exposure relative to the development of disease, and the appropriate exposure metric (that is, what aspect of EMF exposure actually exerted a biological effect?).

Epidemiologists applied themselves to addressing these issues, and over a period of twenty years the quality of the studies improved substantially. The assumption underlying this research was that following the logic and methods of the discipline should lead to a clarification of the relationship of EMF with cancer. This clarification might take one of three forms. First, a stronger association of EMF with disease might emerge from better studies. Second, improved studies might show no evidence of a relationship. Third, some other factor might be identified that would explain the initial association of wire codes with cancer. This program occupied epidemiologists working in this area from the 1980s through the late 1990s. However, the tendency to work within the confines of one's discipline and to focus on technical and methodological issues—understandable though it is—can be an occupational hazard. Their inward looking posture made the epidemiologists less disposed to pay attention to questions that lay outside their area of expertise. The possibility that there might be no biological effects from the very weak electromagnetic fields encountered in everyday life was rarely acknowledged.

THE MELATONIN HYPOTHESIS

No one had ever proposed a plausible biological mechanism by which power-frequency electromagnetic fields could cause leukemia or brain cancer. However, starting in 1987, an intriguing hypothesis was put forward that identified a specific mechanism by which electric power might be implicated in a much more common tumor, breast cancer. Richard Stevens, an epidemiologist at Battelle Pacific Northwest Laboratories, proposed that electrification might be associated with increased rates of breast cancer owing to its effects on the hormone melatonin.[18] Stevens cited limited evidence suggesting that two aspects of electric power—exposure to EMF and to artificial light at- night—could decrease the normal nocturnal rise in melatonin. Melatonin, a hormone produced by the pineal gland in response to darkness, has a number of physiologic properties that are relevant to carcinogenesis. First, it inhibits the production of estrogen and other hormones believed to fuel the growth of breast cancer. It also inhibits tumor growth in experimental animals and humans and is a free-radical scavenger. Finally, melatonin has immunostimulatory properties. Thus, suppression of melatonin by exposure to EMF or to light at night might increase the risk of breast cancer either through the deregulation of estrogen or by other means. While human data supporting an effect of EMF exposure and light at night on melatonin levels were scant, many epidemiologists agreed that the hypothesis was worthy of further study. The "melatonin hypothesis" provided epidemiologists with an attractive and seemingly plausible rationale for how by-products of electric power could contribute to the development of breast cancer, and a number of studies were initiated in the 1990s to explore this possibility. Adding to the appeal of the hypothesis was the recognition by the mid-1990s of the failure of large prospective studies to confirm the influential hypothesis that dietary fat intake was an important risk factor for breast cancer. The new hypothesis' focus on electric power also fitted in with the growing interest in the possible role of environmental pollution in the etiology of breast cancer.

THE PHYSICISTS GET INVOLVED

By the early 1990s, a number of physicists had begun to turn their attention to the issue of health effects of extremely low frequency electromagnetic fields. They were concerned about what they saw as the grossly distorted public perception of a threat from ambient exposure to EMF fostered by the media and by regulatory agencies and fed by epidemiologic studies. Furthermore, they felt that valuable resources were being wasted on the study of this dubious threat. The involvement of physicists was to have an important impact on the further course of the controversy.

In the late 1980s, Robert K. Adair, a physicist at Yale University, became interested in the claims regarding health effects from EMF. He had devoted a thirty-year career to research in elementary particle physics, was a member of the National Academy of Sciences, and held the title of Sterling Professor of physics at Yale. As he approached retirement age, he had looked around for a problem he could work on without having to run a large laboratory typical of research in high-energy physics. Through this wife Eleanor Adair, an environmental biologist and authority on microwave radiation, he attended several meetings on the biological effects of EMF and got to know many of the key figures in that area. Adair saw that there was "a bit of a gap" in the physics of biological interactions of low-frequency EMF, which he wanted to understand better. In 1991 he published a major paper in the journal *Physical Review* in which he used fundamental physical principles to call into question the possibility of health hazards from ambient exposure to power-frequency EMF.[19] Entitled "Constraints on biological effects of weak extremely-low-frequency electromagnetic fields," the paper started out from the position that, due to the low energies of such fields, any contribution to cancer or leukemia incidence could not be due to the breaking of bonds in DNA. Rather, it would have to involve "less catastrophic effects" that are not well characterized or understood.

Adair defined weak fields as electric fields below 300 V/m in air (the mean electric field at the Earth's surface is about 100 V/m) and magnetic fields no greater than 50 μT (or 500 mG), the strength of the Earth's magnetic field. In order for an externally generated electric or magnetic field to have an effect on cellular physiology, the fields would have to exceed the level of endogenous "thermal noise." As he put it to me, all of the molecules in our body are "jiggling around" at a temperature of 36.7°C. This normal level of thermal noise is referred to as kT, that is, a constant times temperature in degrees Kelvin. If the mean energy of the molecules in one's body is altered by as little as 3 percent, "then you're going to be dead." In his 1991 paper, Adair demonstrated that due to the resistivity of tissues and cell membranes, fields actually penetrating the body are in fact far weaker than the thermal noise effects—by many orders of magnitude—and therefore cannot be expected to have any significant effect on the biological activities of cells. Having described the properties of thermal noise within tissues and cell membranes, Adair proceeded to consider different aspects of EMF—electric fields, static and changing magnetic fields, pulsed magnetic fields, and different types of resonance—demonstrating for each scenario that the fields produced are well below the level of thermal noise. In each case he invoked the principles of classical physics to bring clarity to the discussion of specific mechanisms by which EMF could affect biology. He concluded that "there are good reasons to believe that weak ELF fields can have no significant biological effect at the cell level—and no strong reason to believe otherwise."

In spite of the paper's thoroughness and provocative conclusion, Adair was not certain that he was right. As he put it in an e-mail to me in 2004:

> Negative arguments, such as those that say that weak fields cannot affect biology are always suspect as being possibly incomplete. I was not completely converted myself for a long time. When I published the 1991 paper, I rather expected that someone would come up with an angle that I had missed. Indeed, it was much later, after I had considered matters at greater length—and published perhaps ten more papers on the subject—that I became a complete convert myself to the view that environmental fields of 10 mG indicted in the epidemiology cannot possibly affect biology. By that time, no one had found any convincing theoretical process—or convincing experimental data—that would allow power-line frequency fields less than 500 mG to affect biology. Since any effects near threshold must increase with the square of the magnetic field (for AC fields, the average field strength is always zero) or the energy density, the safety factor is not 500/10, or 50, but 2500.[20]

In an article written for the layman in the late 1990s and entitled "The fear of weak electromagnetic fields," Adair was both more outspoken and more accessible.[21] He likened concern over weak EMF from power lines to the fear that leaves falling from trees could fracture a person's skull. "Electric fields alleged to be carcinogenic and generated in humans by the 60 Hz, 5 milligauss (mG) magnetic fields from an electric power distribution system will be only about ten millionths of a volt per meter (V/m) and cannot induce an energy transfer to biologically significant molecules greater than one-millionth kT." "Direct magnetic effects are also possible. . . . But at 60 reversals a second, the magnetic forces cancel out and the energies transmitted to magnetic elements in animals by 60 Hz, 5 milligauss, fields can be expected to be less than 1/10,000 kT. Neither birds, bees, fishes, nor humans can even detect such weak 60 Hz fields, let alone be harmed by them." How then was one to account for the numerous experimental studies purporting to show effects of exposure to EMF? Adair pointed out that after more than twenty years of research, there were no reproducible, agreed upon effects, and he attributed most findings to experimental error. He contends that much of the research in this area, fed by public concern and the availability of funding, was published and accepted uncritically by many who had a stake in there being health effects of ambient EMF. As an experimentalist in high-energy physics, Adair judged the vast majority of experimental work on this question to be of extremely poor quality.[22] The reporting of an exciting new finding was typically followed by a failure of other researchers be able to reproduce it. To give just two examples: a key experiment by one prominent researcher was judged to involve scientific misconduct;[23] another piece of work by the Environmental Protection Agency's lead EMF researcher was shown by Adair to be explainable only if the data were somehow corrupted.[24] If Adair is right, the failure of reviewers and editors to assess research in this area with the requisite degree of rigor and skepticism has contributed to a distorted perception of the issue.

A colleague of Adair's at Yale, William R. Bennett Jr., had also used basic physical principles to clarify the EMF issue. Bennett holds professorships in both engineering and physics and is the coinventor of the gas laser. He had served on the Oak Ridge Associated Universities panel, which produced the highly critical report in 1992. Bennett's involvement in the controversy over potential human health effects of EMF led him to write a book on the topic, as well as an article published in *Physics Today*.[25] The article gives a carefully reasoned account of how EMF from man-made sources compare to natural fields present in the environment and in our bodies. Having undertaken a survey of electric and magnetic fields typically encountered in the urban environment, Bennett determined that the highest sustained ELF fields were associated with electric railroads. As a "worse-case scenario," he calculated the fields that would be experienced by a person standing barefoot on the wet tracks of an electric railroad. Although the external electric field near the barefoot railroad walker's head would be 12,000 V/m, the peak internal field would be only about 80 μV/m (or 1/10,000,000th the external field). This last value is many orders of magnitude smaller than the electric fields normally occurring within the body.

Another difficulty confronting epidemiologic studies of EMF was raised by Adair in a paper published in the *Proceedings of the National Academy of Sciences* in 1994.[26] For the most part, epidemiologic studies had assumed that the pertinent exposure metric was the mean field strength. Adair commented that this assumption "seems to have been made by default with no guidance from biophysical models." In actuality, he argued, biological responses to weak 60-Hz electric and magnetic fields must vary as the square of the field strength. This argument directly addressed the question acknowledged by epidemiologists that in investigating the effects of EMF they were operating in the dark as to what *metric*—that is, what parameter of EMF—was pertinent in inducing biological effects. If Adair was right, the parameterization of exposure in the epidemiologic studies was wide of the mark, and their elaborate statistical analyses might be meaningless. It is symptomatic of the disjunction between the epidemiologists and the physicists raising fundamental questions about the biophysical effects of EMF that this paper was not cited by the epidemiologists.

The vocal case mounted by Adair, Bennett, and other physicists for consideration of fundamental biophysical properties in the evaluation of the health effects of EMF was instrumental in persuading the American Physical Society and the National Research Council of the National Academy of Sciences to undertake independent reviews of what was known about the health effects of EMF. The American Physical Society's report was published in 1995 and the National Research Council report in 1997.[27] Both documents came to the overall conclusion that there was no compelling evidence of adverse health effects on humans.

However, these reports from prestigious institutions did put an end to the controversy. When the National Research Council report was published, three members of the panel, all former presidents of the Bioelectromagnetics Society, put out their own press release to counter that of the National Research Council. Furthermore, the tension between skeptics and those who felt that, although weak, the data could not be completely dismissed left its imprint on the report itself. The report's executive summary was written by the chairman of the committee, Charles Stevens, a neurobiologist at the Salk Institute, and presents a forceful conclusion that the existing science provided no evidence of a hazard from exposure to 60-Hz EMF:

> Based on a comprehensive evaluation of published studies relating to the effects of power frequency electric and magnetic fields on cells, tissues, and organisms (including humans), the conclusion of the committee is that the current body of evidence does not show that exposure to these fields presents a human health hazard. Specifically, no conclusive and consistent evidence shows that exposures to residential electric and magnetic fields produce cancer, adverse neurobehavioral effects, or reproductive and developmental effects.[28]

If one reads the body of the report, and particularly the section on the epidemiology, the message is quite different, leaving open the possibility of effects.[29] However, whatever the tensions within the committee, most people only read the executive summary, and it was the overall conclusions that got reported in the media. The American Physical Society and the National Research Council reports represented important milestones in putting the EMF problem in a scientific perspective.

A NEW GENERATION OF STUDIES

By the early to mid-1990s the emphasis of epidemiologists, like Savitz, on improving the quality of epidemiologic studies of EMF began to bear fruit. In 1991 a study from Los Angeles reported results on 232 cases of childhood leukemia and an equal number of control children.[30] In contrast to previous studies, assessment of exposure in this study included a 24-hour measurement in the child's bedroom and a number of spot measurements in other locations within the house and outdoors in addition to wire coding. Measurements were made in up to two homes lived in during relevant "etiologic period" between the estimated date of conception and a date preceding diagnosis. Although wiring configuration was associated with leukemia, with a doubling in the odds ratio for very high relative to very low current and underground configuration combined, the researchers found no association of measured fields with leukemia risk.

A major study of magnetic fields and cancer in children residing near Swedish high-voltage power lines was published in 1993 by Feychting and Ahlbom.[31] The study base consisted of all children under the age of 16 years who had lived on a property within 300 m of any of the 220 and 400 KV power lines in Sweden between 1960 and 1985. Using the Swedish Cancer Registry, a total of 142 cancer cases were identified, including 39 cases of leukemia and 33 central nervous system tumors. A total of 558 healthy children were selected from the base population for comparison. Exposure was assessed by spot measurements and by calculation of the magnetic fields generated by the power lines, taking distance, line configuration, and historical information on the loads into account. When historical calculations were used, a nearly threefold excess risk of leukemia was found for children with higher exposure, whereas no association was seen for central nervous system tumors, lymphomas, or all childhood cancers combined, contradicting the earlier study carried out in Stockholm.

This study had a number of strengths, including the availability of national statistics on historical power loads and a nationwide cancer registry. The design made use of the fact that residents near transmission lines would have significant exposure compared to those living at a greater distance. Of great importance was the fact that selection of controls from the same base population as the cases greatly reduced the likelihood of selection bias, which was a concern in studies carried out in the United States.

Feychting and Ahlbom interpreted their results as "providing more support for an association between magnetic fields and childhood leukemia than against it." However, in spite of making use of the entire Swedish population under age 16 over a twenty-five year time period, the number of cancers of specific types was limited. Also, as was pointed out in a letter to the *American Journal of Epidemiology*, the reported association with leukemia was one of a very large number of comparisons made by the researchers and hence could well have arisen by chance.[32] Furthermore, as the writers of the letter pointed out, one of the study's key results was not statistically significant, to which Feychting and Ahlbom replied that they never claimed that it was.

In July 1997 the results of a large case-control study of childhood leukemia carried out by the National Cancer Institute appeared in the prestigious *New England Journal of Medicine*.[33] This study was considerably larger and methodologically more rigorous than most of the previous studies. The researchers enrolled 638 children with acute lymphocytic leukemia (ALL) who were under 15 years old and 620 matched controls. Exposure to EMF was assessed by means of a 24-hour measurement in each child's bedroom, spot measurements in other rooms and near the front door, and wire coding (conducted in a blinded fashion) of the subjects' current and former homes. Using the measurement data, the researchers computed a time-weighted average residential magnetic field. They reported that risk of ALL was not increased in children who had resided in homes in the highest wire code category or in homes with higher average measured fields (0.20 μT or greater) compared with <0.065 μT. Furthermore, there was no suggestion of an increased risk with increasing level of exposure. The main results are shown in the table 4.3.

Publication of the National Cancer Institute childhood leukemia study and the National Research Council report represented serious challenges to the notion that power-frequency EMFs posed a cancer hazard. It is fair to say that these two reports, following the American Physical Society report, marked a turning point in the widespread, uncritical acceptance of EMF as serious threat to the nation's health. However, not all parties were willing to conclude that there was no compelling evidence of a hazard. The final

TABLE 4.3 Risk of Childhood Leukemia (ALL) by Time-Weighted Average Summary Levels of Residential 60-Hz Magnetic Fields

Magnetic Field Level (μT)	No. of Cases	No. of Controls	OR (95% CI)
<0.065	206	215	1.00
0.065–0.099	92	98	0.96 (0.65–1.40)
0.100–0.199	107	106	1.15 (0.79–1.65)
≥0.200	58	44	1.53 (0.91–2.56)
0.200–0.299	29	26	1.31 (0.68–2.51)
0.300–0.399	14	11	1.46 (0.61–3.50)
0.400–0.499	10	2	6.41 (1.30–31.73)
≥0.500	5	5	1.01 (0.26–3.99)

Source: Linet et al., 1997.

report from the National Institute for Environmental Health Sciences EMF-RAPID program, which was published in 1999, stuck a somewhat different note.[34] If a tension was discernable in the National Research Council report between the overall conclusions announced in the executive summary and the section on the epidemiology, in the NIEHS report this tension emerged into the conclusions themselves. While acknowledging that virtually all of the experimental and mechanistic studies failed to demonstrate adverse effects and that the epidemiologic evidence was weak and inconsistent, a majority of the NIEHS working group, nevertheless, on the basis of the admittedly weak and inconsistent epidemiologic evidence, voted to classify EMF as a "possible carcinogen." The Director of NIEHS, Kenneth Olden, recommended that, in view of the remaining uncertainty concerning health effects due to EMF, passive regulatory action aimed at reducing exposures was warranted. But in the key paragraph of the report summarizing the epidemiologic evidence, the reasoning is strained and highly questionable:

> None of the individual epidemiologic studies provides convincing evidence linking magnetic field exposure with childhood leukemia. Hence, in making an assessment, one must rely upon the evaluation of the data as a whole using expert judgment and the meta-analyses as a guide. The pattern of response, for some methods of measuring exposure, suggests a weak association between increasing exposure and increasing risk. The small number of cases in these studies makes it impossible to firmly demonstrate this association. This level of evidence, while weak, is still sufficient to warrant limited concern.[35]

First, one should immediately be wary if the data are not compelling without resorting to "expert judgment" and meta-analysis. "Expert judgment" seems too suggestive of experts who have an investment in a field to which they have devoted years of work. Furthermore, the results of meta-analysis can be highly questionable when the studies being combined differ in the details of their methods and the biases that affect their results. It is hard to escape the impression that the reluctance of the NIEHS working group to close the door on the possibility of EMF as a cause of leukemia had more to do with its members' stake in this area of research than with scientific rigor.[36]

The null results of the National Cancer Institute childhood leukemia and several other large studies in the late 1990s led epidemiologists to carry out further analyses of the existing data to determine whether distorting factors, such as confounding and selection bias, could explain the association between EMF and childhood leukemia.[37] Although the results of studies differed, taken as a whole, they suggested that selection bias and confounding by such factors as socioeconomic status and traffic density could partially account for the observed association.

In addition, two pooled analyses of existing studies of EMF and childhood leukemia came out in 2000, in effect summing up twenty years worth of epidemiologic research on EMF and childhood leukemia.[38,39] Both analyses involved going back to the original data from each of the studies in order to make full use of the available information, rather than relying solely on the published reports. By using the original data, the researchers were able to maximize the comparability of different studies carried out in different geographic locations. Furthermore, pooling studies greatly increased the sample size and made it possible to examine the dose-response relationship. It also made it possible to look at differences between studies (rather than assuming homogeneity amongst the studies) as well as interrelationships between different exposure measures (i.e., wire codes and measured fields). All in all, these two papers provide some of the most thoughtful discussions of the actual data from the epidemiologic studies that have appeared to date.

There was substantial overlap in the studies included in the two analyses, so that many of the same investigators are authors on both papers. However, the two analyses included somewhat different numbers of subjects and used somewhat different approaches, which could account for some of the divergences in their results. Both analyses failed to find clear evidence of a dose-response relationship between field strength and risk of childhood leukemia. In both studies, there was no indication of increased risk below the highest exposure category (>0.3 μT in Greenland et al. and >0.4 μT in Ahlbom et al.), but both reported a modest but significantly increased risk for the highest category (i.e., roughly a twofold risk in both studies). This high exposure group, however, represented only a minute fraction of the total population (2 and 0.8 percent, respectively). Both analyses yielded results that contradicted the so-called "wire code paradox," but their findings differed. (The wire code paradox refers to the fact that use of wire codes—a crude surrogate—tended to show stronger associations with childhood cancer than use of actual measurements of fields.) Greenland and colleagues found that "very high" versus "very low" wire code was associated with increased risk and that wire codes were "strongly" associated with measured fields. (They noted, however, that these results were based on only four studies and were very unstable.) In contrast, Ahlbom et al. found no increased risk associated with residing in homes in high wire code categories. In fact, they noted that "the measured fields are low in all the wire-code categories." This represented a striking reversal of the association of wire codes with measured fields, which had so preoccupied epidemiologists throughout the 1980s and 1990s, and cast doubt on the usefulness of wire codes. The authors went on to say, "The reasons for the elevated risk estimates for high wire-code categories in the earlier North American studies are unclear, although considerable potential for bias has been noted for both studies carried out in Denver." (The difference between the two analyses regarding wire codes may be explained in part by the fact that Ahlbom et al. excluded the study by Savitz et al., whereas Greenland et al. included it.)

Both papers emphasized the methodological difficulties of epidemiologic studies of EMF, given the incomplete assessment of exposure, the potential for selection bias and confounding and the fact that the only hint of increased risk occurs in the miniscule group with the highest exposure. Greenland et al. were unsparing in articulating the uncertainty of the results of their analysis:

> Biases due to measurement errors are undoubtedly present in and vary across all of the studies, but their assessment is not wholly straightforward. . . . Only under fairly restrictive conditions can one be certain that the net bias due to such error will be toward the null. Unfortunately, there is little or no evidence to establish such detailed attributes of the errors, and there is no basis for assuming such attributes are the same across studies and measures. . . . Furthermore, the associations are imperfect enough to indicate that probably all of the measures suffer considerable error as proxies for any biologically relevant exposure measure (if one exists). . . . These problems should further expand the considerable uncertainty apparent in our results.[40]

In their conclusion, Ahlbom et al. devoted a forceful paragraph to the resoundingly null evidence from laboratory studies, implicitly correlating those null findings with the lack of an association in 99 percent of the children included in the pooled analysis:

> The results of numerous animal experiments and laboratory studies examining biological effects of magnetic fields have produced no evidence to support an aetiologic role of magnetic fields in leukaemogenesis (Portier and Wolfe, 1998). Four lifetime exposure experiments have produced no evidence that magnetic fields, even at exposure levels as high as 2000 μT, are involved in the development of lymphopoietic malignancies. Several rodent experiments, designed to detect promotional effects of magnetic fields on the incidence of leukaemia or lymphoma have also been uniformly negative. There are no reproducible laboratory findings demonstrating biological effects of magnetic fields below 100 μT.[41]

Noting that the elevation in risk observed in 0.8 percent of their data (the twofold increase in the risk of leukemia) needs to be explained, they suggested that selection bias may have accounted for part of the excess.

Both papers emphasized that, in order to clarify the nature of the association observed in the small group with the highest measured exposure, future studies would have to be conducted in highly exposed populations, preferably in densely populated countries like Japan. These two analyses of the existing epidemiologic evidence conveyed a very different picture from that found in the majority of epidemiologic papers and reviews (with notable exceptions like Poole and Trichopoulos).

EMF AND BREAST CANCER

In 2002 and 2003, three large studies appeared, doing for breast cancer what the National Cancer Institute study had done for childhood leukemia.[42] The studies had been carried out in Seattle, Los Angeles, and on Long Island and incorporated at least one twenty-four hour measurement in the woman's bedroom in addition to spot measurements in other rooms and wire coding. (I should note that I was a coauthor on the Long Island report.) These studies took measurements in multiple homes or limited their enrollment to residences that had been occupied by the study subjects for at least fifteen years prior to entry into the study. In this way, they attempted to assess EMF exposure relevant to the decade or two preceding the diagnosis of breast cancer. They also evaluated effects of "harmonics" (i.e., multiples of the 60-Hz fields) in addition to the fundamental 60-Hz frequency fields, ground currents, and electric appliances, including electric blankets. Strikingly, none of these studies showed any hint of an association of exposure to EMF with breast cancer, whether wire codes or actual measured fields were examined. Carried out in different areas of the country and published within a short interval, these studies went a long way toward ruling out the possibility that EMF was an important cause of breast cancer.

In addition, the study from Seattle, Washington, had measured nighttime urinary melatonin levels over a period of three days in two different seasons of the year but found no strong or consistent relationship between residential exposure to EMF and melatonin levels. Since the melatonin hypothesis had been proposed in the late 1980s, most studies designed to address the question had provided little support for an effect of EMF on melatonin levels in humans.[43] (Studies of exposure to light at night and shift work, which can depress the normal rise in melatonin during the night, in relation to breast cancer have yielded inconclusive results. However, there is intriguing evidence that blindness may reduce the risk of breast cancer.) Finally, eight studies that examined the association between use of electric blankets and breast cancer also failed to find evidence of increased risk. These studies contributed important information since electric blankets, especially older ones, entailed relatively high exposures. In publishing the eighth null report on the association of electric blanket use and breast cancer, the editor of the journal *Epidemiology* voiced the opinion that the issue was now dead.[44]

LESSONS OF THE EMF STORY

The evolution of the EMF controversy over a period of nearly twenty-five years is marked by a number of paradoxes and contradictions. By failing to present a balanced and critical view—such as that presented by Poole and Trichopoulos in 1991—epidemiologists and experimentalists engaged in research on EMF contributed to an exaggerated concern about health effects from power lines and other sources. Their work fed the journalistic and regulatory appetites for "evidence" of this hazard. At the same time, the high level of anxiety among the public, federal agencies, and industry was responsible for the allocation of research funds that would not otherwise have been forthcoming. As a result of intensive work by epidemiologists, however, a marked improvement in the quality of studies took place, and the later studies were able to either dispel concern about a hazard (in the case of female breast cancer) or put a very low upper limit on the possibility of a hazard (in the case of childhood leukemia). It is clear that these studies would not have been undertaken on such a large scale or as quickly as they were had it not been for the enormous societal anxiety and economic interests at stake. Work got done that otherwise would not have been done. So, one might reason that, in spite of the cost and the widespread and unnecessary public alarm, something positive came out of the process. But it is reasonable to ask, what is the value of that work? Not surprisingly, the answer to this question depends on whom you ask.

One response was given by Edward Campion, an editor at the *New England Journal of Medicine*, who wrote an editorial accompanying the 1997 National Cancer Institute childhood leukemia study. Under the title "Power lines, cancer, and fear," Campion summed up the EMF issue in the bluntest possible terms.[45] After enumerating the weaknesses of the epidemiologic studies, he wrote: "Moreover, all these epidemiologic studies have been conducted in pursuit of a cause of cancer for which there is no plausible biologic basis. There is no convincing evidence that exposure to electromagnetic fields causes cancer in animals, and electromagnetic fields have no reproducible biologic effects at all, except at strengths that are far beyond those ever found in people's homes." (It is noteworthy that, in addition to the National Research Council report, Campion cited the 1991 paper by Poole and Trichopoulos and the Oak Ridge report of 1992.) He went on to conclude that, owing to unfounded fears about EMF, hundreds of millions of dollars were wasted on investigating a hypothesis that never had any biological evidence to support it and that made no contribution to advancing our understanding of the causes of childhood leukemia.

A very different point of view is implicit in a commentary by Savitz entitled "Health effects of electric and magnetic fields: Are we done yet?" which appeared in the journal *Epidemiology* in 2003.[46] Where Campion addressed the issue in the broadest terms and took an outspoken position, Savitz, writing for epidemiologists, confined himself to the question of what further research after the past twenty-five years was worth undertaking and publishing on this topic. Even allowing for the fact that Savitz's commentary is directed at epidemiologists, the narrowness of his focus is striking. The whole implication of the piece is that epidemiologic research on EMF has basically accomplished what it can accomplish and that, barring a major improvement in methods or the emergence of solid "empirical" evidence that would justify the study of specific outcomes, new studies are unlikely to be productive. While stressing the need for empirical, as opposed to merely theoretical support for a hypothesized biological mechanism, Savitz does not see fit to comment on the fact that decades of empirical research on mechanisms whereby EMF could affect biology have failed to produce any replicable findings. In contrast to Campion, he does not refer to skeptics like Poole and Trichopoulos or the Oak Ridge report, to say nothing of physicists like Adair and Bennett. While drawing distinctions about what further epidemiologic work on EMF is warranted, Savitz chooses not to draw any lessons from the past twenty-five years or to even point out any problems that are visible in retrospect. His narrow focus allows him to avoid confronting questions concerning the value of the work that was done, why it essentially failed to uncover evidence of a hazard, and whether the issue could have been approached differently from the start.

While Campion's editorial may have performed a useful function in questioning the value of research on EMF, in one respect his position is extreme. Even physicists like Adair and Bennett and skeptical epidemiologists like Poole and Trichopoulos do not go so far as to deny any value to research on this topic. Rather, they argue that the facts of the case did not justify a "crash program" and special "set-aside" funding for such studies. In their view, studies of health effects of ELF-EMF should have competed for funding in the general grant pool on an equal footing with other investigator-initiated proposals. Adair also concedes that some research may be justifiable, including studies of high exposures, such as those encountered by users of electric blankets.

There is no need to be quite as categorical as Campion is. One can take exception to the exploitation of poorly founded fears and still concede that useful knowledge has been gained through the epidemiologic studies. Some of those studies have been able to address other potential risk factors such as overcrowding and traffic congestion, and it is conceivable that these lines of work could lead to new insights in the future. The fact that high measured fields in the recent pooled analyses are associated with a slight increased risk of childhood leukemia points to something that still needs clarification.

A further justification for conducting such studies, given by Savitz and others, is that this research can allay serious public concerns about novel exposures. As Savitz pointed out to me, epidemiological studies cannot rule out the existence of a very weak effect, but if they are well designed and large enough, they can put an upper limit on a potential hazard. They can allow us to say, "Look, we know that there are no dire effects from everyday exposure to EMF. This is not some catastrophe."

Thus, one benefit of continued study of EMF is that epidemiologists were able to put a potential hazard into a sharper perspective. In a review article published in 1995, the Swedish epidemiologists Feychting and Ahlbom calculated that if residence near high-voltage transmission lines in Sweden was in fact associated with a doubling of the risk of childhood leukemia (as indicated by three high-quality studies from Scandinavia), this would translate to *less than one case per year out of a total of 70* being due to this exposure.[47] Given the many methodological issues surrounding the epidemiology and the considerations raised by the critics, this estimate of 1/70 can be looked at as the sort of upper bound referred to by Savitz. However, in response to a question posed to Adair by Louis Slesin, the editor of *Microwave News*, Adair put his own estimate of the probability of any contribution of ELF-fields of less than 10 milligauss to cancer at one in a million.

In looking back at the trajectory of the EMF controversy over a period of twenty-five years, it is significant that epidemiologists involved in studying the effects of EMF rarely made use of the criteria for judging the causality of an association, first proposed by Austin Bradford Hill and discussed in chapter 2.[48] How does the available evidence fit the four most important of Hill's criteria? First, regarding the magnitude of the association, the Wertheimer and Leeper study reported risk estimates of 2–3, but most subsequent studies reported risk estimates between 1 and 2. Risk estimates in this range are sufficiently weak that the effects of bias are difficult to rule out. Second, as we have seen, there were numerous inconsistencies when the results of the studies were examined in detail. Third, there was also a striking absence of a dose-response relationship with increasing strength of measured fields. Finally, the criterion of "biological plausibility" was never satisfactorily addressed by those making the case for adverse health effects of EMF. This criterion requires epidemiologists to take account of what is known from other disciplines. In the case of EMF, this would have meant a rigorous assessment of the laboratory and mechanistic studies *and* an understanding of the relevant physics.

If physicists like Adair and Bennett were right, there were compelling reasons, based on the biophysics of the interaction of ELF-EMF with human tissues, to think that no mechanism was likely to be discovered in the future. Their arguments shed light on why many intriguing experimental results—seized on by Brodeur and highlighted in the newsletter *Microwave News*—had never been convincingly reproduced. The case put forward by the physicists was directly relevant to addressing biological plausibility. Allowing for a time lag of several years (since the articles of Adair and Bennett were not published in epidemiology journals), one might have expected that by the mid or late 1990s their work would have been cited by epidemiologists working on EMF. However, this was not the case. The careful arguments made by Adair and Bennett concerning the possibility of adverse—or beneficial—health effects from weak ELF-EMF encountered in everyday life went without mention in the epidemiologic literature. This is even more surprising given that Adair had attended meetings of the Bioelectromagnetics Society and published in the society's journal *Bioelectromagnetics*, as did many epidemiologists working on EMF. One cannot escape the impression that epidemiologists caught up in the methods and data from their own discipline had no time or interest to enlarge their field of vision to incorporate the highly relevant point of view of the physicists. In epidemiology papers published in the years 2000 to 2004, I was able to find only one reference to a paper of Adair's[49] and no references to his most wide-ranging and provocative paper from 1991, "Constraints on biological effects of weak extremely-low-frequency electromagnetic fields" or to Bennett's 1994 paper in *Physics Today*. Furthermore, in no article on the epidemiology of EMF that I am aware of has there been any discussion of the viewpoint put forward by the physicists.

In his interview with me, Adair, who, at nearly eighty, is mild mannered and laughs easily, expressed the view that the epidemiologists were arrogant, that they felt that their discipline was paramount because they were engaged in measuring actual exposure and linking it to the occurrence of disease in humans. What is clear is that the epidemiologists' "tunnel vision" kept them focused narrowly on the results of their studies to the exclusion of fundamental insights about the phenomenon whose effects they were investigating. After all, nature does not recognize the distinctions between academic disciplines.

Box 4.1 Major Take-Home Points

The question of possible adverse health effects of exposure to ambient electromagnetic fields (EMF) from power lines, home appliances, and other sources became an issue of interest to scientists and of concern to the public with the publication of an epidemiologic study in 1979. Although relying on a crude assessment of exposure, this study seemed to point to an association between childhood cancer and proximity to power distribution lines.

Further epidemiologic studies were carried out, and, by the late 1980s, the topic of EMF had become of focus of intense regulatory and media attention and enormous public concern. The quality of the epidemiologic and laboratory studies, however, was poor, and the results were weak and inconsistent. Nevertheless, the apparent indications of links to a growing number of diseases created the widespread perception of a hazard.

There was a split among scientists, with those engaged directly in research on EMF feeling that there was justification for further study of this question, and "outsiders" arguing that decades of research had produced no evidence of reproducible biological effects of the weak electromagnetic fields encountered in daily life. Furthermore, a number of physicists argued a priori that these fields were highly unlikely to have any effects because they were many orders of magnitude smaller than the energy levels within the body.

Owing to the publication in the late 1990s, and after, of a number of large epidemiologic studies showing no association of EMF exposure with either childhood leukemia or female breast cancer, as well as the acknowledgment by regulators of the failure of experimental studies to yield reproducible results, fears of EMF began to wane and to be put in perspective.

5

THE SCIENCE AND POLITICS OF RESIDENTIAL RADON

One might have imagined that radon, issuing innocently as it does from the ground, would be difficult to politicize.

—Robert Proctor, 1995

We've created a statistical illness, multiplying a very small risk by very large populations to come up with the frightening figures.

—Ernest Létourneau, 1987

Radon is a colorless and odorless radioactive gas resulting from the decay of radium and ultimately from uranium, which is ubiquitous in the Earth's crust and occurs in varying amounts in soil and rock. Normally, radon dissipates harmlessly in the air, but it can seep into homes through fissures and openings in the foundation and can accumulate to produce high levels in homes. Although radon itself is inert, it undergoes radioactive decay to polonium-218 and polonium-214 with the emission of high-energy alpha particles. These radon "progeny" or "daughters" are electrically charged and can attach to fine dust particles in the air, enabling them to be inhaled and deposited in the lining of the lungs. There they can undergo further decay emitting alpha radiation that can damage cells in the lung, leading to the development of cancer. Due to the very short range of alpha particles, radon can only affect the cells lining the lung with which it comes in direct contact, and radon is not known to cause cancer in organs other than the lung.

Geological formations in certain regions of the country have high uranium content, and homes in these regions are likely to have high radon levels. Yet even within the same geographic area, the radon concentration in homes can vary by several orders of magnitude,[1] so that it is not possible to predict whether a given house has a problem without making direct measurements. This is because the radon level in a given home is dependent on soil permeability, building construction, air pressure within the house, as well as other factors, such as barometric pressure and temperature. However, other things being equal, houses in areas with high uranium content are more likely to have high residential radon levels than houses in other areas. In addition to its presence in the air, radon can also be present in water, but only about 5 percent of exposure in the United States is believed to derive from exposure to water through activities like showering and cooking.[2]

Extensive evidence for the carcinogenicity of radon comes from studies of underground miners exposed to extremely high levels and from animal and molecular studies.[3] However, the effects of exposure at the much lower levels typical of exposure in homes are unknown. Furthermore, the question of the effects of domestic radon exposure is fraught with complexities, paradoxes, and imponderables that were often lost sight of when radon became a focus of intense public and government concern in the mid-1980s. The gap between what the science indicated and the message that was orchestrated by the federal government, politicians, and the media is perhaps nowhere greater than on the subject of radon. In addition, there is disagreement within the scientific community about the extent of the hazard to the general population from exposure to radon in homes and elsewhere in daily life. However, not all points of view have been given equal weight in the regulatory and political arenas and in the media. Consequently, there has been a slanting of the science, and we will have to examine how this takes place.

FROM OCCUPATIONAL CARCINOGEN TO DOMESTIC THREAT

Although radon itself was not discovered until 1898, the earliest observations of its effects on health date from the sixteenth-century physician Georgius Agricola, who noted the frequent occurrence of a mysterious "mountain sickness" (*Bergkrankheit*) among silver miners in the Erz Mountains straddling present-day Germany and the Czech Republic. Agricola observed that some women were widowed as many as seven times due to the miners' high mortality rate. In 1879, based on autopsies and clinical examinations, the physicians F. H. Harting and W. Hesse identified the miners' malady as pulmonary malignancy,[4] and by the early twentieth century the condition was demonstrated to be primary cancer of the lung.[5] Over the next thirty years, additional reports suggested that roughly half of the Erz miners eventually succumbed to lung cancer.[6] Following the discovery of radioactivity in the late 1890s and the subsequent recognition of radiation-induced cancer, by the 1930s it was hypothesized that high levels of radon gas emanating from the uranium-rich ores in the mines was the culprit. However, this explanation was not universally accepted, and some physicians attributed the miners' high rate of lung cancer to inbreeding within the population or to other compounds present in the mines.

It was only with the increased demand for uranium following World War II and the expansion of uranium mining that radon was firmly linked to the increased rates of lung cancer among miners. In the 1950s the Public Health Service and the Colorado State Department of Health began sampling radon progeny levels in mines on the Colorado Plateau, and later the first epidemiologic cohort study of Colorado Plateau uranium miners was initiated.[7] Analyses of the mortality experience of this cohort, which first appeared in the 1960s and 1970s, demonstrated that miners suffered from an excess of lung cancer and that the risk of disease increased with increasing exposure.[8] The results of the Colorado Plateau study were supported by the findings from other cohorts of miners in North America, Europe, and China, and collectively these studies provided the major source of data on the long-term effects of heavy exposure to radon progeny.

Attention to radon as a potential health hazard in homes first arose in the 1970s from the identification of "hot spots" like Grand Junction, Colorado, and central Florida, where homes had been built on the uranium-rich "tailings" left over from the milling of uranium and phospho-gypsum. (Tailings are the huge quantities of fine, grey radioactive sand left over from the milling uranium ore to produce "yellowcake," which after further refining can be used to fuel nuclear reactors or in making nuclear weapons.) It was also recognized, based on the Swedish experience, that certain building materials, such as alum shale, could contain significant amounts of uranium, producing elevated radon levels in homes. In response to these findings, both the Department of Energy and the Environmental Protection Agency (EPA) established programs to evaluate radon as an "indoor air pollutant." Prompted by concern about a hazard to homeowners living above mill tailings, Congress passed the Uranium Mill Tailings Radiation Control Act in 1978, and, to comply with the act, in 1983, the EPA set a mandatory "action level" of 4 pCi/L of air as a goal, with an upper limit of 6 pCi, for these homes.[9] [The two units most commonly used to measure radon levels in homes are picocuries per liter of air and becquerels per cubic meter (Bq/m^3). A picocurie is a trillionth of a curie; 1 Ci is the amount of radiation emitted by 1 g of uranium. One picocurie per liter of air is equivalent to 37 Bq/m^3]. As pointed out by several observers, it is ironic that the agency was well aware that many homes had "natural" radon levels that exceeded those found in homes built over tailings but made no effort to address this more common situation.[10]

In spite of the focus on high radon levels arising from landfills contaminated with radioactive wastes, before the mid-1980s, there were indications that the problem might be of much wider scope. During the 1970s and early 1980s, scientists at Lawrence Berkeley Laboratories and Argonne National Laboratories conducting research into the effects of energy conservation standards on indoor air quality were surprised to find high radon levels in some model homes.[11] And following the accidental release of radioactive material from the Three Mile Island nuclear power plant in 1979, measurements in nearby homes showed the presence of high radon levels (20–100 pCi) that were not due to the accident.[12]

While these developments were of great interest to health physicists and epidemiologists concerned with the biological effects of ionizing radiation, there was no widespread concern about radon in homes and, hence, no pressure for a national policy regarding radon. This was to change as a result of the fortuitous discovery of extraordinarily high radon levels in a home in Boyertown, in eastern Pennsylvania. In December 1984, Stanley Watras, an engineer working on the construction of Philadelphia Electric Company's Limerick nuclear power plant outside of Pottstown, Pennsylvania, set off radiation detectors on his way into the plant. The source of his contamination was traced to his home, which had levels of 2,800 pCi/L, or 700 times higher than the recommended standard.[13] Boyertown is located near the southwestern end of the Reading Prong, a geologic formation rich in uranium ore that stretches from eastern Pennsylvania through north central New Jersey and into southern New York. In the months following the Watras incident, a survey of more than 1,600 homes in Berks County found that nearly 40 percent had elevated radon levels.[14]

One of the many paradoxes surrounding the question of the effects of exposure to indoor radon is that radon contamination of homes, except for rare instances involving contaminated landfills, is due to naturally occurring uranium in soil and rock and thus did not fit the pattern of man-made industrial pollution. For this reason, it did not clearly come under the EPA's jurisdiction, and the agency was slow to set policy regarding radon. According to Richard Guimond, who became head of EPA's newly formed radon division in 1986, the Watras incident was responsible for galvanizing the agency into action.[15] In May 1985, on the basis of very limited survey data, the EPA estimated that one million homes around the country might be contaminated at levels posing a serious lung cancer hazard and that from 5,000 to 20,000 lung cancer deaths per year could be caused by residential radon.[16] However, by mid-1986 the agency had increased its estimate of the number of affected homes to 8 million,[17] and it later revised its estimate of the number of lung cancer deaths attributable to radon to 7,000 to 30,000.[18]

Federal and state officials, politicians, and the media were quick to characterize the newly discovered problem in the most extreme terms. A front-page article in the May 19, 1985, edition of the *New York Times* carried the headline "Major Peril is Declared by U.S.," while an *Atlanta Constitution* story was titled "Radon in Homes Could Kill 30,000 Yearly."[19] *Newsweek* reported the EPA's estimate of the number of affected homes under the headline "Radon Gas: A Deadly Threat—A Natural Hazard is Seeping into 8 Million Homes."[20] According to Robert Yuhnke, the Environmental Defense Fund's regional counsel in Denver who had been active on radon issues, the situation in the Reading Prong area of Pennsylvania represented the "early stages of a cancer time-bomb waiting to go off."[21] EPA labeled radon "probably the biggest public health problem we have,"[22] and the Assistant Surgeon General of the Public Health Service, Vernon Houk, termed radon-induced lung cancer "one of today's most serious public health issues."[23] Media coverage devoted to radon gas in homes surged in the mid-1980s following the publicity surrounding the Watras home and the ensuing attention to testing of homes. For example, the number of articles on indoor radon published in the *New York Times* went from zero in 1983 to twenty-two in 1985, to forty-seven in 1986, and to forty in 1987, after which it began to taper off.[24]

Although EPA officials acknowledged that there was virtually no epidemiologic evidence available regarding the health effects of exposure to radon at residential levels (as opposed to data on exposure to much higher levels in the confined and dusty environment of mines), the agency proceeded to make strong claims about the threat from radon in homes—claims that often had little scientific basis. In addition, the EPA chose to downplay certain aspects of the radon issue, and this led to a national radon policy that many scientists and other observers judged to be misguided. In keeping with its philosophy that there is no safe level of exposure to a carcinogen, the EPA adopted a "conservative" approach to radon. This led it to set what some radon authorities considered to be an overly stringent threshold for remediation of 4 pCi/L of air and to recommend further reducing levels to 2 pCi/L.

The administrators and scientists at EPA who shaped the agency's radon policy undoubtedly believed that their guidelines were justified by what the agency considered to be a major public health threat. EPA officials argued that indoor radon posed a much more serious hazard than other substances that it had regulated in the past, such as formaldehyde and asbestos. However, a number of factors in the political landscape appear to have played an important role in shaping the agency's radon policy. First, radon, which had been a localized issue limited to communities contaminated with radioactive debris, quickly became an issue of national concern following the Watras incident. Throughout the second half of the 1980s, a number of powerful members of Congress took up the issue and made radon the focus of congressional hearings. These included Senator Frank Lautenberg of New Jersey (whose home happened to be in Montclair, the site of a bungled and protracted clean-up of a radon-contaminated residential landfill), Senator George Mitchell of Maine, and Representatives Edward Markey and Henry Waxman—the last was chairman of the House Subcommittee on Health and the Environment. Many of the most vocal politicians urging government action on radon were Democrats and liberals, and the radon question quickly took on a political coloration. Early on, the Reagan administration was open to the charge of not addressing the radon problem with sufficient vigor, and politicians like Lautenberg and Waxman exerted powerful political pressure on the government to act aggressively to control radon in homes, schools, and other buildings.[25] EPA responded to this pressure by taking a hard line on radon, and, in return, the agency was given sole responsibility for articulating a national radon policy.[26] Second, it has been suggested that EPA's radon policy was also influenced by the agency's need to distinguish itself from the Department of Energy, which had a long-standing radon program but which, owing in part to its roots in the nuclear power industry, took a more cautious and less activist approach, acknowledging the many uncertainties pertaining to radon.[27] Finally, it has been pointed out that, in the antiregulatory climate prevailing during the Reagan years, radon provided the administration with the perfect opportunity to demonstrate its concern for public health without having to confront industry, since the culprit in this case was Mother Nature. Some have characterized the EPA's radon campaign as "environmentalism on the cheap."[28]

THE EPA'S RADON CAMPAIGN

In August 1986 the EPA published two widely circulated brochures spelling out the main elements of its radon policy: *Radon Reduction Methods: A Homeowner's Guide* and *Citizen's Guide to Radon: What It Is and What to Do About It.*[29] The first publication acknowledged that techniques for remediating home radon levels were poorly understood, but it provided guidelines for those with a serious problem. The second booklet was less tempered and made no attempt to put the radon issue in perspective or to acknowledge the considerable uncertainties. On its first page it stated, without elaboration, that "scientists estimate that from about 5,000 to about 20,000 lung cancer deaths a year in the United States may be attributed to radon."[30] It went on to note that roughly 85 percent of the 130,000 expected lung cancer deaths in the United States in 1986 were attributable to cigarette smoking. But no explicit mention was made of the phenomenon of synergism between smoking and radon exposure, which means that most of the lung cancer deaths ascribed to radon would not have occurred in the absence of smoking. There is only a vague statement on the penultimate page (page 12) of the brochure to the effect that "smoking may increase the risk of exposure to radon." However, the off-hand nature of the statement and its relegation to the end of the pamphlet suggested that it was an afterthought of little importance. Thus, the brochure conveys the erroneous impression that radon must be responsible for a substantial proportion of the remaining 15 percent of lung cancer that is not explained by smoking.

The *Citizen's Guide* also included a chart designed to convey the magnitude of the lung cancer risk due to different levels of home radon in terms of equivalence to the risks from smoking and exposure to chest X-rays. For example, a level of 15 pCi/L of air was equated to the risk of lung cancer of a pack-a-day smoker. The clear message from this chart is that even low levels of indoor radon are something to be concerned about. But these supposed equivalences are meaningless because they fail to distinguish between the effects of radon in smokers and never smokers.

The cornerstone of the EPA's radon policy was the setting of a recommended "action level" for homes of 4 pCi of radon per liter of air, with the further recommendation that, if possible, indoor levels should be reduced to 2 pCi. In order to identify homes exceeding 4 pCi, the agency urged that all homes below the third floor be tested, since there was no valid method of predicting which homes had a problem. Beyond the decision to set a relatively stringent action level, in its pronouncements on radon the EPA tended to overstate the number of homes with high levels, as well as the magnitude of the risk associated with a given radon exposure. In doing so, it created a great deal of needless anxiety and confusion both about the extent of a health hazard and about the need for testing and the benefits of testing.

Responding to increased concern about radon, in 1986, Congress passed the Radon Gas and Indoor Air Quality Research Act, which essentially made the EPA responsible for determining the magnitude of the indoor radon hazard and the appropriate response. It is significant that the Department of Energy, which had a long-standing research program on radon, was not even mentioned in the bill.[31] Two years later, Congress enacted a more ambitious bill, the Indoor Radon Abatement Act, which further delineated EPA's responsibility to provide updated information regarding the health risk posed by radon and techniques for testing and mitigation, as well providing assistance to state and regional radon programs.[32] But, without doubt, the act's most striking provision, contained in its opening paragraph, was its espousal of the "national long-term goal of the United States" of reducing radon levels in buildings to the level of outdoor air.[33] (The median radon level in outdoor air is about 0.5 pCi/L). While such a goal was far from being technologically feasible and would have been prohibitively expensive, supporters argued that it represented a valid long-term objective.

In spite of its wide distribution (one million copies were printed), the *Citizen's Guide to Radon* was unsuccessful in motivating large numbers of homeowners to test for radon, and the EPA soon escalated its efforts to gain the public's attention.[34] In the fall of 1988, appearing on national television, EPA Administrator Lee Thomas claimed that up to one third of U.S. homes had radon concentrations exceeding the agency's action level of 4 pCi/L of air.[35] The agency also equated the lung cancer risk due to a daily radon exposure of 4 pCi to that from smoking up to half a pack of cigarettes per day, although there was no scientific basis for this equivalence.[36] Again, as pointed out by Philip Abelson, a nuclear physicist and editor of the journal *Science*, what was most inexcusable about these estimates was the failure to distinguish between radon's effects in smokers and nonsmokers.[37]

In 1989, in collaboration with the Ad Council, a private, nonprofit organization, the EPA developed an aggressive campaign to overcome the public's apathy regarding radon, producing advertisements for print media and billboards, as well as spot television commercials. These materials featured anxiety-provoking images and alarming "statistics" to convey a sense of urgency about the lethality of the insidious, odorless gas. Many of the ads urged viewers or readers to call a hotline, 1-800-SOS-RADON, to obtain further information. One pamphlet was titled "Protect your family against Radon . . . the silent killer." It included an image of a chest

X-ray with the caption, "Having Radon in your home is like exposing your family to hundreds of chest X-rays yearly."[38] Many of the ads prominently featured the claim that radon is the "second leading cause of lung cancer in the United States—after cigarette smoking."[39] While technically correct, as we shall see, this formulation is incomplete and quite misleading.

THE PARADOXES OF RESIDENTIAL RADON

The paradoxes and uncertainties surrounding the question of the health effects of low-level radon exposure made it a challenge for scientists and regulators to present an accurate assessment of the risk to the general population. There was always the danger that certain aspects of the problem would be overstated and become the focus of attention and that other equally important aspects would be neglected. Risk assessments conducted by the National Council for Radiation Protection and Measurements and the National Research Council, which produced estimates of the number of lung cancer deaths caused by indoor radon, were careful to spell out the many qualifications and uncertainties attaching to their estimates. However, it was the number of estimated deaths—and, usually, when a range was given, the higher number—that was latched onto by the media, politicians, and the EPA, and where the estimates came from, and what the necessary qualifications and attendant uncertainties were, were lost sight of. Even in scientific papers the perspective could be skewed. But ignoring the paradoxes and the uncertainties only ensured that the resulting policy would be distorted and misguided. For this reason, it is worth pausing to summarize a number of the central paradoxes or contradictions surrounding the question of low-level radon exposure.

The fact that radon did not fit the profile of industrial environmental pollutants has already been noted. A second paradox—some would call it an irony—is that although radon is estimated to account for 55 percent of the average person's exposure to ionizing radiation,[40] other less important sources of radiation, such as nuclear power plants and food irradiation loomed much larger as concerns for most people.[41] Though quantitatively more important, apparently, radon is less frightening because it is "natural." In spite of the EPA's aggressive ad campaign to overcome what it perceived to be the public's apathy toward radon, only a small percentage (approximately 5 percent) of homeowners have been motivated to measure the levels in their home.[42] The public's perception of the radon hazard had its own logic and determinants, but prominent among these determinants were the confusing and incomplete public information that was disseminated by the EPA.

Without doubt, the greatest paradox of residential radon is the fact that the vast majority of lung cancer cases ascribable to radon occur in smokers (current and former). In an interview with the *New York Times* in January 1988, the chairman of the National Academy of Sciences' Committee on the Biological Effects of Ionizing Radiation, Jacob Fabrikant, a professor of radiology and biophysics at the University of California, was quoted as saying, "Radon risk is primarily a risk of smokers unless, of course, nonsmokers spend time in a house with extremely high levels of radon. This is another reason to end cigarette smoking."[43] This clear and forceful formulation by a respected scientist should have been kept firmly in view in all discussions of the radon issue since it has profound implications both for understanding the radon issue and for formulating policy. First, it implies that the most effective way to reduce radon-induced lung cancer in the population is to motivate remaining smokers to quit. In those who have already quit, lung cancer risk declines over a number of decades, eventually approaching the risk of those who never smoked. A corollary of the overwhelming importance of smoking as the dominant cause of lung cancer is the fact that lung cancer occurring in lifetime nonsmokers is so rare that it is difficult to determine how much of a risk, if any, average levels of domestic radon pose to those who have never smoked. As we will see, this is a central problem for studies of residential radon.

The interaction between radon and smoking is so fundamental that to speak of radon without making this a central take-home message is inevitably to distort the issue. Yet, the fact that most lung cancer cases due to radon exposure are likely to occur in smokers has not gotten the attention it deserves. It is revealing that in a number of key articles, the narrow focus on radon left little room for adequate attention to, or even mention, of this central fact.[44] In spite of efforts to place the radon problem in its proper perspective by keeping the much greater effect of smoking in view, all too often the linkage was lost sight of, and radon became a thing-in-itself, divorced from its crucial context. In this way, the hazard was easily misconstrued and exaggerated.

Another paradox entails the fact that, based on its extrapolation from the miner data, the National Research Council estimated that, if all homes with elevated radon levels were reduced to 4 pCi/L of air, this would only prevent one third of radon-induced lung cancer deaths. In other words, fully two thirds of these deaths were estimated to occur in homes with radon levels below the EPA "action level." The reason for this is that the vast majority of the population has exposures in this lower range, and the lower level of exposure applied to the much larger "population at risk" contributes more deaths. However, again, it needs to be remembered that approximately 90 percent of radon-induced lung cancer deaths will occur in smokers. Furthermore, although rarely mentioned, it stands to reason that if two thirds of radon-induced lung cancer deaths occur in those with average residential exposures below 4 pCi, an even greater percentage of these lung cancers (that is, greater than 90 percent) are likely to occur in smokers. This is because, at very high exposures, radon is more likely to cause lung cancer in the absence of smoking.

CRITICISM OF THE EPA CAMPAIGN

To many in the scientific community, including health physicists who had devoted their careers to the study of radon, the EPA's radon policy was not only objectionable on scientific grounds because it ignored many of the complexities and uncertainties pertaining to radon, but it was also bad policy. The agency's estimate of the number of homes exceeding the action level of 4 pCi/L of air was overstated due to the fact that the early surveys were not systematic and tended to oversample regions with high radon levels and due to reliance on short-term screening measurements that were often made in basements, where radon levels are usually higher than in the lived-in areas of the home. Critics also pointed out that the action level of 4 pCi was set too low, thereby defining a much larger number of homes as requiring remediation. Both Sweden and Finland had adopted well-designed radon programs based on sound science that had met with public acceptance, using action levels of 10 and 20 pCi/L, respectively.[45] Canada had relaxed its cutoff from 4 to 20 pCi.

Largely on the basis of the 1986 analysis of available surveys of radon in homes conducted by Anthony Nero of Lawrence Berkeley Laboratory, the EPA estimated that 12 percent, or 8 million, of the nation's seventy million homes exceeded 4 pCi/L. However, according to Nero, 7 percent, or five million, would have been a more accurate figure. And he argued that efforts should be focused on identifying the much smaller number of homes (about 1 percent) with average radon levels exceeding 10 pCi and remediating those homes.[46]

Others, like the physicist and Nobel laureate Rosalind Yalow of the Bronx Veterans Administration Hospital, pointed out that the fact that the rate of lung cancer in never smokers had remained low throughout the twentieth century put a very low limit on any effect of radon in the non-smoking population.[47] It could also be argued that, because of the large increase over the course of the twentieth century in the proportion of the population living in apartment buildings, as opposed to single-family homes, the radon exposure of the population may actually have decreased throughout the past century. Such "ecologic" considerations and broad trends cannot resolve the issue of the effects of residential radon exposure, but they merit consideration and can help to put the issue in perspective.

In addition to its estimates of the total number of lung cancer deaths attributable to radon, the EPA also presented a range of risk estimates at various exposure levels.[48] According to its figures, people spending 75 percent of their time for 70 years in a home with an average indoor level of 4 pCi/L were said to face an extra risk of dying of lung cancer of between 1.5 and 5 in a hundred. Those with an exposure level of 10 pCi/L would incur an excess risk of 3 to 12 in a hundred. At an exposure level of 200 pCi/L the projected additional lung cancer risk is shown to be between 44 and 77 in a hundred. Actually, these figures are meaningless because it is not specified whether they apply to smokers or nonsmokers, as EPA eventually did in later publications. Naomi Harley, a leading radiation expert at New York University School of Medicine, who assessed radon risks for the National Council on Radiation Protection and Measurements in the early 1980s, called the EPA estimates "outlandishly high."[49] According to her, a one-in-a-hundred excess lung cancer death rate at an exposure of 4 pCi/L should be the midpoint, not the lower bound of the range. Regarding the EPA estimates at the highest levels ranging from 44 to 77 percent for those with an average exposure of 200 pCi, Harley noted that even among the nineteenth-century miners, who had the greatest radon exposures ever recorded, fewer than half died of lung cancer. She concluded that the highest EPA estimates were "out of line with reality."[50]

Another source of the EPA's overestimates of the effects of radon was pointed out by Kenneth Warner and colleagues at the University of Michigan School of Public Health.[51] They noted that the EPA's estimates of the effects of home exposure to radon were severely overstated because the agency failed to take account of residential mobility. They cited the statistic that on average people in the United States change residences 10 to 11 times throughout their lives. Only homes that exceed the EPA's action level of 4 pCi/L of air are targeted for remediation, and such homes represent only about 7 percent of the nation's housing stock. Given that the average indoor radon level is 1.25 pCi, the odds are greatly in favor that a move from a high radon house will be to a lower radon house. For this reason, calculating lifetime risk due to an individual's "*current* exposure level is not generally a good guide to their *cumulative* lifetime exposure; it is the latter that determines lung cancer risk." "In particular, for people living in the high-radon homes the EPA targets for action, normal patterns of residential mobility mean that the vast majority will experience cumulative lifetime exposures equivalent to residing in homes having, on average, much lower radon levels." Furthermore, the authors went on to demonstrate by use of examples that mitigation of all homes exceeding 4 pCi would reduce the typical individual's risk by "no more than 30% percent and usually much less."

Perhaps the most comprehensive and incisive critique of the EPA's radon policy came from two specialists in environmental engineering, William Nazaroff of the University of California at Berkeley and Kevin Teichman of the EPA itself. Writing in the journal *Environmental Science and Technology* in 1990, they described what was known about the radon problem and the likely costs and benefits of the current government policy.[52] Their central argument was that, when the apparent synergy between smoking and radon exposure and population mobility are taken into account, the cost of remediation of indoor radon "may be less attractive to individuals than to society as a whole." As an illustration, they showed that for a typical family of four lifetime nonsmokers who live in a house for ten years before moving, the cost of reducing the mean indoor radon level from 260 Bq/m^3, or 7 pCi/L, (the mean level of houses exceeding the 150 Bq/m^3 action level) to 150 Bq/m^3, or 4 pCi, would be about $3,000. The corresponding reduction in the lifetime risk of one lung cancer death in the family would be 5 percent in relative terms, or about 1 in a 1,000 in absolute terms. For a family of smokers, the reduction in absolute risk is more than ten times greater, but only one lung cancer death would be averted among 60 such households, yielding a cost of $180,000 per premature death averted. Regarding implementation of the 1988 Indoor Radon Abatement Act's goal of reducing indoor radon to the level of "ambient air outside of buildings," Nazaroff and Teichman determined that, "even if this were technically feasible," the costs would be "staggering"—on the order of one trillion dollars.

As an alternative to EPA policy, like Anthony Nero, Nazaroff and Teichman proposed identifying the very small fraction of the housing stock (approximately 70,000 homes) with average radon levels greater than 800 Bq/m^3, or 22 pCi. Occupants of such homes, they wrote, have an extraordinarily high lifetime risk of developing lung cancer, which should not be ignored. Because these homes tend to be clustered geographically, they can be identified more efficiently by taking into account geological information, thus avoiding the necessity of measuring every household. A focus on highly contaminated homes should be combined with efforts to improve radon measurement and mitigation techniques, foster a public consensus, and set realistic goals for dealing with the problem of domestic radon. Such an approach, Nazaroff and Teichman commented, would yield benefits that cannot be achieved "if the radon problem is treated as an epidemic that requires rapid countermeasures with little regard for scientific uncertainties and accompanying costs."[53]

Finally, a number of well-informed observers criticized the agency's motivational campaign to persuade homeowners to test their homes as gratuitously alarmist and fundamentally wrong-headed.[54] Philip Abelson and Anthony Nero blasted the EPA for its "terror" campaign designed to overcome public apathy. Abelson charged that "one of the weaknesses of the EPA is that it seems unable to learn. Its basic policies were set nearly twenty years ago. Whenever a risk is identified, the EPA takes what it calls a conservative approach. This entails developing worst-case scenarios and giving credence to sloppy data if they indicate a greater risk. Experiments that later show that no risk exists are disregarded. Very rarely indeed has the EPA loosened regulations on the basis of new, valid scientific data."[55]

Over the past twenty-five years, the question of the health effects of radon has received a prodigious amount of attention from scientists, professional bodies, and government agencies. Progress in delineating the hazard posed by residential radon has been incremental, and the picture that has emerged is complex, with many gaps and uncertainties, but nevertheless with certain fixed and relatively well-established points. However, within the scientific community there are differing assessments of the importance of residential radon as a health hazard, and the consensus view—the dominant interpretation as well as the message that has been put out to the public—has emphasized certain aspects of the problem and downplayed others. Even some of the most authoritative and comprehensive reports are characterized by a peculiar ambivalence and ambiguity, which has contributed to an exaggeration of radon as a public health problem. For this reason, we must now turn to an examination of some of the major documents on the health effects of domestic radon over the past twenty years and how they have been interpreted for a wider audience.

STUDIES OF UNDERGROUND MINERS

When the issue of radon in homes arose in the mid-1980s, studies of underground miners constituted the vast majority of the available evidence regarding the effects of radon exposure on humans. Reports on the mortality experience of uranium miners in the Colorado Plateau had appeared starting in the 1960s and indicated that miners experienced elevated lung cancer risks.[56] In 1984, at the behest of the U.S. EPA and the Nuclear Regulatory Commission, the National Research Council of the National Academy of Sciences undertook a detailed report on the health effects of alpha-emitting radionuclides, including radon and its progeny. The report was the fourth report of the National Research Council Committee on the Biological Effects of Ionizing Radiation, or BEIR, and is referred to as BEIR IV. The specific charge to the committee was to analyze the miner data and extrapolate effects downward from the heavy exposures encountered in underground mines to the much lower levels of radon in the domestic situation. Three years in the making, the report, entitled *Health Risks from Radon and Other Internally Deposited Alpha-Emitters*, came out in January, 1988, immediately following the period of greatest publicity surrounding domestic radon.

BEIR IV

The core of the 1988 report consists of an analysis of the association of radon exposure and lung cancer deaths in four cohorts of underground miners and the extrapolation of risk estimates to the general population due to environmental radon exposure. The committee obtained the original data from the four cohort studies and used specially developed statistical methods to reanalyze it. In the risk model selected by the BEIR IV Committee as best describing the relationship of radon exposure and subsequent lung cancer risk among miners, the relative risk was found to increase with increasing exposure but to decrease with time since exposure and with increasing age. Radon concentrations in underground mines are expressed in terms of "working levels" (WL). (A working level is any combination of radon progeny in one liter of air that results in the emission of 1.3×10^5 million electron volts of alpha particle energy.) Exposure in the mines is expressed in terms of "working level months" (WLM), which is equivalent to one working level for a working month (170 hours). The committee's risk projections based on the miner data to the levels pertinent to the general population yielded an estimate that lifetime exposure to 1 WLM/year increases the risk of lung cancer by a factor of about 1.5 over the current rate for both males and females, given the prevalence of smoking at the time (mid-1980s). (The average environmental exposure is equivalent to 0.2 WLM/year.[57]) Excess lifetime lung cancer mortality due to radon was estimated at 350 lung cancer deaths per million person-WLM/year.[58] Following its estimates, the committee added a crucial sentence: "In all these cases, most of the increased risk is in smokers in whom the risk is 10 or more times greater than that in nonsmokers."[59] The report went on to list a number of sources of uncertainty affecting its estimates, acknowledging that "the committee recognizes that the differences between the risks in mining and domestic environments and the interaction between smoking and exposure to radon progeny remain incompletely resolved."[60]

Although the committee acknowledged the uncertainties attaching to its estimates, some of the limitations of the available data may still have been understated. Of the four miner cohorts, only one (the Colorado Plateau cohort) had any information about smoking habits. Furthermore, of the 151 lung cancer cases used in the analysis, we are not given the number occurring in lifetime nonsmokers but are told that only 22 of the 151 consumed between 0 and 4 cigarettes per day.[61] Thus, the BEIR IV analysis could say little about the effects of radon exposure in never smokers, a key question concerning the effects of radon on the general population. Furthermore, inspection of the relevant table[62] of the report indicates that the number of smokers in the individual cells, particularly at lower levels of radon exposure and lower levels of smoking, are so sparse as to limit the ability to describe the interaction between radon exposure and cigarette smoking. As Richard Hornung, a radon expert at the National Institute of Occupational Safety and Health put it in 1991, "we have virtually no data on low-level exposed miners who didn't smoke."[63]

BEIR VI

Eleven years after the publication of the BEIR IV report, in 1999 a second report from the National Research Council entitled *Health Effects of Exposure to Radon*, provided a new, updated analysis of the miner data, this time based on 11 cohorts and including 68,000 miners and 2,700 cases of lung cancer, as compared to the 4 cohorts and 360 lung cancer deaths in BEIR IV.[64] The large increase in the available data permitted a refinement of the model developed in the earlier report.

After extensive analysis, BEIR VI concluded, on the basis of what was known from studies of underground miners and an understanding of genomic damage caused by alpha particles, that exposure to residential radon is "expected to be a cause of lung cancer in the general population."[65]

The analysis further indicated that there was a "synergistic effect" of radon together with smoking, although the precise nature of the interaction—whether "multiplicative" or "sub-multiplicative"—could not be determined. Most importantly, the committee estimated that between 15,400 and 21,800 lung cancer deaths per year in the United States could be attributed to domestic radon exposure, although the authors indicated that, given the uncertainties in the analysis, the actual number of deaths could be between 3,000 and 33,000. (The authors also estimated that between 2,100 and 2,900 of the lung cancer deaths caused by radon occur in never smokers.) These numbers, the committee noted, make radon "the second leading cause of lung cancer" after cigarette smoking.[66]

For all the sophisticated statistical modeling provided in the report, two major problems cast doubt on the ability make accurate estimates of the effects of radon at residential levels based on the miner data. These are the dearth of data on miners exposed at low radon levels and the powerful effect of smoking.

The core of the report consists of two components: (1) the extrapolation of the miner data from the 11 cohorts down to the levels encountered in homes and (2) the meta-analysis of the available case-control studies of residential radon. Each of these topics was addressed in a paper in the *Journal of the National Cancer Institute* published in the mid-1990s,[67] and the results were presented in greater detail and in updated form in the 1999 report. I will refer both to the papers and to the report.

The analysis of the combined miner data from the 11 cohorts showed that cumulative radon exposure in miners was directly related to risk of lung cancer—in a linear manner—in both ever smokers and never smokers (fig. 5.1). In addition, based on theoretical considerations, the committee posited that there was no threshold, i.e., no level below which radon would not affect the risk of lung cancer, although it acknowledged that the existence of a threshold could not be ruled out. Application of the "linear no threshold model" to the miner data implied that even at the low levels encountered in most homes, radon exposure would carry some risk.[68]

In addition to analyzing the full range of the miner data (ranging from 0 to 2500 WLM of cumulative exposure), the committee carried out a separate analysis, restricted to cumulative exposures below 50 WLM. This level was judged to represent the upper range of residential exposure. (Living in a house with a radon level of 1 pCi/L of air for 1 year is roughly equal to 0.2 WLM of exposure. In order to accumulate 50 WLM of exposure in a house, a resident would have live in a house with a radon level of 10.8 pCi/L for 25 years).

In fig. 5.1, domestic exposure, in the range below 50 WLM, corresponds to the leftmost point on both graphs. This level of exposure has a relative risk that appears to be indistinguishable from 1.0 (i.e., no increased risk).

Of the eleven cohort studies analyzed, only six had any information on the smoking history of the miners, and even where there was information

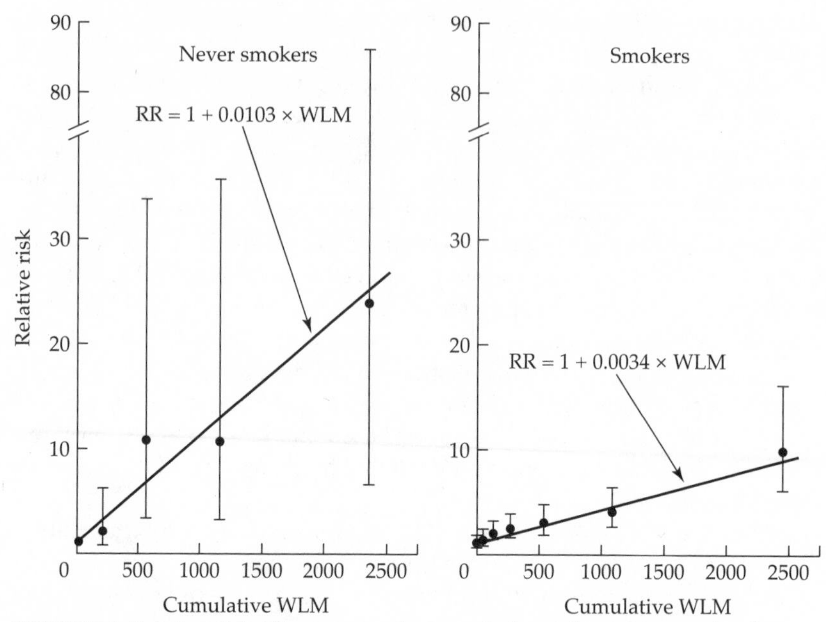

FIGURE 5.1 Relative risk of lung cancer by cumulative radon concentration for all eleven miner cohort studies combined, for reported never smokers and ever smokers. *Source*: Lubin et al., 1995.

in some cases this was limited or was only available on a portion of the subjects. On the basis of this imperfect information, 58 percent of the miners appear to have smoked. However—and this critical fact is not directly stated—*only 3 percent of lung cancer deaths in those cohorts with information on smoking history occurred in never smokers*. [The 3 percent figure is obtained by dividing the number of lung cancer cases occurring among miners who never smoked (64—obtained from BEIR VI, Table C-13) by 1,814—the total number of lung cancer cases in the 6 miner cohorts that had some information on smoking history (extracted from the tables in Appendix D, BEIR VI, p. 254 ff.).] What this means is that at the very low exposure levels typical of homes, the number of lung cancer cases in miners who never smoked must be extremely small. (Lubin and Boice note that there are 358 lung cancer deaths among miners with cumulative exposure below 50 WLM.[69] If 3 percent of these are never smokers, that equates to about 11 never smokers.) In other words, there is little basis for assessing the effects of radon exposure in this range on never smokers. Furthermore, it is difficult to see how it is possible to assess the interaction between radon and smoking at low levels of exposure since this depends on having a credible estimate of the effect of radon in never smokers.

Based on a meta-analysis of the eight case-control studies published as of that time, the committee concluded that the trend in the relative risk for the combined studies was statistically significant and indicated a relative risk of 1.14 (95 percent confidence interval 1.0–1.3 at 150 Bq/m^3, or 4 pCi.[70] Figure 5.2 shows the results of the meta-analysis of the indoor studies (black squares) and those from the underground miner studies (open squares) restricted to exposures below 50 WLM. Both sets of estimates are in the same range, and the committee concluded that the results of the case-control studies were consistent with those seen in miners. However, the five risk estimates (open squares) based on the miner data at low exposure levels (below 50 WLM) are all between 1.0 and 1.5, and none is statistically significantly different from 1.0 (no increased risk), reflecting the paucity of data. The pooled residential data (black squares) show a small elevation in the relative risk for the two highest exposure levels, but these data include smokers.

In addition, there are a number of other problems and uncertainties acknowledged by the committee that will be mentioned in the next section. Thus, for the committee to make projections about the number of lung cancer deaths attributable to domestic radon exposure in never smokers based on the miner data appears questionable. Nevertheless, these estimates are included in the report's Executive Summary and are widely accepted as

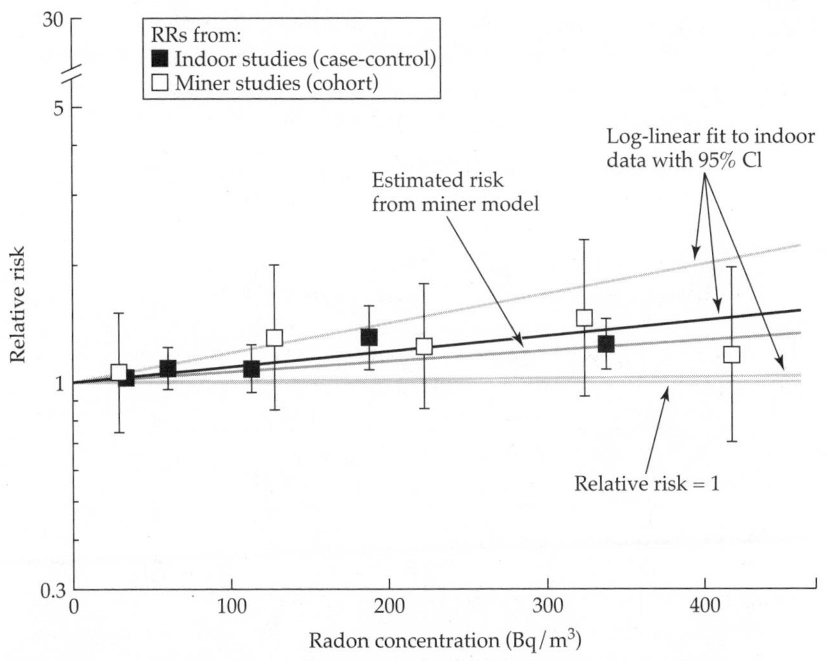

FIGURE 5.2 Summary relative risks (RR) from meta-analysis of indoor-radon studies and RRs from pooled analysis of underground miner studies, restricted to exposures under 50 WLM. *Source*: National Academy of Sciences, 1999.

fact, even though the many necessary qualifications enumerated in the report rarely get mentioned.

Both the *Journal of the National Cancer Institute* papers and the BEIR VI report perform a fascinating balancing act. On the one hand, they present the results of careful statistical modeling of the available data and lay out the many limitations and attendant uncertainties pertaining to the analysis. On the other hand, in spite of the many intractable limitations and uncertainties, the authors feel compelled to present estimates of the effects of radon on the general population and in never smokers, and the underlying message is that radon is important as an environmental carcinogen. A reflection of this conceptual balancing act is an ambivalence and ambiguity that permeates these texts. For example, the abstract of the 1995 *Journal of the National Cancer Institute* paper concludes that, "In the United States, 10% of all lung cancer deaths might be due to indoor radon exposure, 11% of lung cancer deaths in smokers, and 39% of lung cancer deaths in never-smokers."[71] The final sentence adds the qualification: "These estimates should be interpreted with caution, because concomitant exposures of miners to agents such as arsenic or diesel exhaust may modify the radon effect and, when considered together with other differences between homes and mines, might reduce the generalizability of findings in miners."[72] Similarly, in the discussion section of the 1997 paper devoted to the meta-analysis of eight case-control studies of residential radon, the authors repeat no less than four times that the "results must be cautiously interpreted."[73] Finally, the closing sentences of the Executive Summary of the BEIR VI report are a consummate feat of balancing complex and competing aspects of the radon problem, as they shift back and forth between different emphases:

> The qualitative and quantitative uncertainty analyses indicated that the actual number of radon-attributable lung cancer deaths could be either greater or lower than the committee's central estimates. This uncertainty did not change the committee's view that indoor radon should be considered as a cause of lung cancer in the general population that is amenable to reduction. However, the attributable risk for smoking, the leading cause of lung cancer, is far greater than for radon, the second leading cause. Lung cancer in the general population and in miners is related to both risk factors and is amenable to prevention.[74]

In this way, the authors of these reports manage to have it both ways. But, in spite of the carefully calibrated balance, in the final analysis, the message conveyed by the reports is that *radon is important and that by itself it accounts for a significant proportion of lung cancer incidence*. In the end, the focus is narrowed to radon alone, and estimates are given for the number of lung cancer cases "due to radon." In order to frame broad conclusions, the carefully articulated caveats formulated by the committee needed to be shunted to the side. In the final analysis, statistical models, which include many uncertainties and assumptions, are used to make estimates based on sparse data, and these estimates are then put forward as valid descriptions of reality.

EXTRAPOLATING FROM MINERS TO DOMESTIC RADON EXPOSURE

As the BEIR VI Committee was careful to acknowledge, differences between radon exposure in underground mines and in homes were numerous and substantial and raised questions about the validity of extrapolating from the former to the latter. First, as mentioned above, levels in underground mines were orders of magnitude higher than the levels in homes. According to Lubin and colleagues,[75] the lowest levels in mines were 50–100 times higher than the average level in homes of about 1 pCi/L of air. There were also important differences in the conditions of exposure in the two environments. As described by Krewski and colleagues,[76] these include the relative proportion of radon to its decay products (which affects how much energy is delivered to the lung), a person's breathing rate (which affects the rate at which radon and its progeny are inhaled and deposited in the lung), and the size of dust particles in the air (which influences the fraction of radon attached to particles and the depth of penetration and retention within the lung). As mentioned earlier, another limitation of the occupational data on miners is that most miners were smokers. In addition, other substances, some of which are carcinogens, including arsenic, silica, and diesel exhaust, were present in certain mines, complicating the identification of effects of exposure to radon alone. Furthermore, most miners worked for only a few years underground, whereas the relevant period for residential exposure is many decades.[77] Finally, the miner data were limited to males and did not address possible differences in exposure and susceptibility of women and children.[78]

It was this awareness of the marked differences between radon exposure in the occupational setting of mines and in homes that provided the impetus for carrying out studies of residential exposure. But, from the outset, radon experts were under no illusions that these studies would be easy to conduct or that they would resolve the question definitively by detecting a strong and unambiguous effect of residential radon.[79]

In his critique of the EPA's radon policy delivered in 1991, Philip Abelson had called for a large epidemiologic study of residential radon exposure as a way to resolve the question that had generated so much alarm and confusion.[80] Such a study would have to be large enough to permit distinguishing the effects of exposure to low levels of radon typical of most homes from the potent effects of smoking. But Abelson argued that the millions of dollars spent on such a study would be a "better investment than spending billions of dollars on remediation that might be a waste of money."[81] Others have argued that the key group in which an effect of domestic radon needed to be demonstrated was lifelong nonsmokers, who accounted for only a small percentage of lung cancer cases.[82] Female never smokers, in particular, should be focused on since historically women have spent more time in the home and typically have less exposure to occupational carcinogens. Furthermore, most lung cancer in never smokers occurs among women. An adequate sized study of this group was projected to require several thousand cases and an equal number of controls.[83] However, a large national study of residential radon as envisaged by Abelson and others was never undertaken. Instead, many small case-control studies were carried out in different areas of the United States and Canada, as well as in Europe and China, and most did not focus exclusively on lifetime nonsmokers. None of these studies came even close to being of a size necessary to tease apart the effects of radon from those of smoking or to detect an effect of radon in lifetime nonsmokers.

CASE-CONTROL STUDIES OF RESIDENTIAL RADON

The cohort study design, which was used to study underground miners, was not feasible to assess the effects of indoor radon on the general population. Such a study would require very large groups of free-living subjects who would have to be followed for a long period of time in order to monitor their exposure and determine their health status.[84] The case-control study offered an alternative that theoretically, at least, could address the question of the effects of residential radon exposure. In this type of study, newly diagnosed cases of lung cancer could be identified through hospitals and cancer registries, and a control group representative of the general population could be identified through random digit dialing, Medicare rosters, or population registries. Cases and controls could then be interviewed concerning their residential history and smoking habits, and radon levels could be measured in the current and former residences.

The major problem confronting case-control studies of radon is that of accurately characterizing an individual's exposure over a period of several decades preceding his or her recruitment into the study. The average American will have lived in as many as eleven different homes during his or her lifetime, and the prospect of measuring even those homes inhabited during the two to three decades preceding recruitment posed formidable problems. Ideally, one would want to obtain long-term measurements (preferably for one year) in each of these homes. However, in practice, this was not feasible because some homes will no longer exist, some will be unoccupied, and in other cases the current owner may refuse to cooperate. Even where homes are accessible, radon measurements made in the present may not reflect past levels owing to modifications in the home or the heating system.

Some studies dealt with this problem by including only cases and controls who had lived in their current home for 10 or more years, while others attempted to obtain measurements in the previous home as well. Still other studies attempted to measure all homes lived in for at least a year within a thirty-year "time window" up to five years before recruitment into the study.

In all studies, there were homes in which measurements could not be made, either because the home was not accessible or because the radon device malfunctioned, and consequently there were gaps in some subjects' exposure history. How one handles such missing data can influence the estimate of a person's cumulative exposure.

Furthermore, in order to accurately characterize an individual's exposure, one would also want to know how much time he or she spent at home during different periods and where, within the house he or she spent time, especially in multilevel homes, because radon levels can vary significantly in different parts of a home and on different floors. Most case-control studies placed one or two radon detectors in the living area and bedroom of residences, but few took measurements on upper floors, and few obtained information on the subject's allocation of time within the house.

The challenges confronting case-control studies of domestic radon were laid out with clarity in a 1990 article entitled "Design issues in epidemiologic studies of indoor exposure to Rn and risk of lung cancer" and published in the journal *Health Physics*.[85] The authors, Jay Lubin, Jonathan Samet, and Clarice Weinberg, were prominent epidemiologists or statisticians in the field of radon. Lubin and Samet were members of the BEIR IV and BEIR VI committees, and Samet chaired the latter. The burden of the article was to caution researchers that, because of the many problems and biases involved in carrying out these studies and particularly in obtaining radon measurements intended to characterize an individual's exposure over many years, studies would have reduced statistical power to test key hypotheses regarding the effects of radon. Specifically, the authors considered the impact of such factors as subject mobility (which tends to reduce lifetime exposure toward the population mean), the choice of the model used to characterize the effects of radon, and errors in calculating exposure. Errors in estimating exposure stem from a number of sources, including errors in the measuring devices, exposure to radon outside the home, "the inability to measure exposures over time in current as well as previous residences, and the unknown relation between measured concentration and lung dose of α energy from the decay of Rn and its progeny." The authors referred to the "formidable methodological problems" that make case-control studies of radon "difficult to carry out and that may limit their interpretation."

Appearing when it did in 1990 before the publication of all but one of the U.S. case-control studies, the paper by Lubin and colleagues provided a much needed caveat concerning what could be expected of individual case-control studies of indoor radon and lung cancer. They wrote that "realistically such studies may never be able to answer many of the subtle questions about risk patterns that burden current risk assessment with uncertainty." And they went on to say:

> Even the most carefully designed and conducted investigations are subject to substantial error in dosimetry, particularly when used to estimate temporally remote exposure. Inappropriate design assumptions with regard to the underlying effect, subject mobility, and exposure distribution also seem inevitable.[86]

In their conclusion, the authors "urge[d] cautious interpretation and reduced expectations for case-control studies." They noted that some studies may fail to show any association simply due to inadequate power owing to insufficient sample size and to biases that reduce the effects toward the null. However, they also cautioned that other studies may indicate an effect that is substantially higher than that expected based on the miner data and that such results should be interpreted carefully since they could be due to small samples and the "selective publication of positive results." The authors looked forward to the future pooling of the many small case-control studies as a way of overcoming the problem of inadequate sample size and obtaining more precise estimates. The results of such pooling efforts have begun to appear and are still in progress.

Although it is a technical paper published in a specialized journal, the points made by the authors are accessible to any interested reader. The paper should have been required reading for anyone interested in understanding the reports of individual case-control studies.

RESULTS OF THE INDIVIDUAL STUDIES

Studies of residential radon exposure using long-term direct measurements began appearing in the early 1990s, and additional studies have continued to appear up to the present. More than twenty studies have been carried out in North America, Europe, and China. Their results are summarized in table 5.1. For each study, the table presents the number of cases and controls, the percentage of never smokers among the cases, the average radon concentration, and the "excess odds ratio" at 100 Bq/m^3. A value above 0 is indicative of increased risk. For example, the excess odds ratio from the Iowa study of 0.44 can be read as a 44 percent increase in the risk of lung cancer associated with an increase in exposure of 100 Bq/m^3.

The studies differ in many respects, including their measurement protocols, the number of homes in which measurements were attempted, and whether they included all cases or were limited to nonsmokers or women. Most studies are suggestive of a small and not statistically significant increase in risk of lung cancer due to radon exposure. Of the 20 studies listed in table 5.1, only four show a statistically significant excess odds ratio associated with radon exposure. Of these four studies, one (conducted in Iowa) shows an excess risk four-and-a-half fold greater than two of the other studies. Two studies—the largest single study carried out in West Germany and a study in Shenyang, China—showed a nonsignificant inverse association between radon levels and lung cancer risk. In other words, increasing radon exposure appeared to be associated with reduced risk of lung cancer. However, no study showed a significant inverse association of radon with lung cancer risk. (A new report on a case-control study conducted in Worcester County, Massachusetts, presents results that are strikingly at variance with those in table 5.1. Thompson and colleagues found that, after controlling for smoking and other potential confounding factors, radon exposure was strongly *inversely* associated with lung cancer risk.[87] What makes this modest-sized study of particular interest is the fact that the investigators attempted to address a number of the weaknesses of previous studies by obtaining detailed information on allocation of time within the house over different life periods and by improving quality control. The results are not included in table 5.1 because the authors did not present the excess odds ratio.)

Interestingly, studies with the highest mean radon concentrations (Czech Republic; Gansu, China; South Finland; France; Stockholm; Iowa; and Winnipeg) did not provide stronger evidence of an effect than the studies where concentrations were lower. The New Jersey study had by far the lowest mean radon concentration in homes but the highest excess odds ratio, whereas the Winnipeg study had a relatively high mean radon concentration but one of the lowest excess odds ratios. In fact, there is no correlation between the radon concentrations and excess odd ratios presented in table 5.1 ($r = -0.008$, $p = 0.97$). This is worth mentioning because regions with higher radon levels should provide a greater opportunity to observe an association due to the wider range of exposures.

In studies that give the smoking status of men and women separately, the proportion of never smokers among male cases ranges from 0.4 percent in England to 5 percent in Gansu, China. Among female cases the proportion of never smokers ranges from 8 percent in Missouri to 88 percent in Gansu, China. Thus, male cases, who comprise the majority of cases, are almost all smokers. This means that, as is true of the miner cohorts, these studies can tell us little about the effect of radon in men who never smoked and are severely limited in their assessment of the interaction between smoking and radon, which requires adequate numbers of never smokers. Even though the proportion of never smokers is higher in female cases, because the number of cases in females is smaller, here too the power to detect an effect of radon exposure in never smokers is limited.

TABLE 5.1 Case-Control Studies of Residential Radon and Lung Cancer

Study	No. of Cases/ Controls[a]	Never Smokers among Cases (%)	Average Radon Concentration (Bq/m^3)[b]	Excess Odds Ratio[c] (95% confidence interval)
New Jersey	480/442 F	14	26	0.56 (-0.22–2.97)
Winnipeg	738/738	3.2	120	0.02 (-0.05–0.25)
Missouri I (nonsmokers)	538/1183		63	0.01 (<0.00–0.42)
Missouri II	512/553 F	8	56	0.27 (-0.12–1.53)
Iowa	413/614 F	13.6	127	**0.44 (0.05–1.59)**
Connecticut	963/949	5.7	33	0.02 (-0.21–0.51)
Utah/south Idaho	511/862	11.0	57	0.03 (-0.20–0.55)
Sweden (Stockholm)	201/378 F	19	128	0.16 (-0.14–0.92)
Sweden (national)	1281/2576	13	107	**0.10 (0.01–0.22)**
South Finland	291/495	1.8	213	0.28 (-0.21–0.78)
Finland (national)	517/517	8.5	96	0.11 (-0.06–0.31)
Southwest England	667/2108 M	0.4	56	0.08 (-0.03–0.20)
	315/1077 F	7.3		
Italy	325/295 M	1.8	96	0.14 (-0.11–0.46)
	59/109 F	39		
East Germany	1046/1414 M	2	74	0.08 (-0.03–0.20)
	146/226 F	26		
West Germany	1214/1865 M	1.9	50	-0.02 (-0.18–0.17)
	235/432 F	31		
Sweden (nonsmokers)	258/487	—	79	0.28 (-0.05–1.05)
France	688/1428	9	128	0.05 (-0.01-0.12)
Czech Republic[d]	210/12,004	17	509	**0.09 (0.02–0.21)**
Shenyang	308/356 F	45	85	-0.05 (<0.00–0.08)
Gansu	563/1232 M	5	223	**0.19 (0.05–0.47)**
	205/427 F	88		

Adapted from Krewski et al., 2005.

[a]Abbreviations: F, females; M, males; NS, never smokers; ExS, ex-smokers.

[b]Values given are mean residential radon concentrations, except for the study from Italy (geometric mean) and the study from Shenyang, China (median).

[c]Excess odds ratio at 100 Bq/m^3. Note that an excess odds ratio greater than 0 indicates potentially increased risk. Bolded numbers indicate statistically significant results.

[d]This study is a cohort study.

POOLING OF INDIVIDUAL STUDIES OF RESIDENTIAL RADON

By 2005, two analyses of the pooled residential radon studies had appeared, one conducted on thirteen European case-control studies, the other on seven North American case-control studies.[88] By virtue of their large numbers, these analyses provided more precise estimates of the risk associated with a given level of domestic radon exposure. And the pooled estimate could then be compared with estimates based on extrapolation from the miner experience. The European analysis included a total of 7,148 lung cancer cases and 14,208 controls, while the North American analysis included 3,662 cases and 4,996 controls. There were 884 lung cancer cases among lifelong nonsmokers in the European analysis and 659 in the North American analysis, enabling a more sensitive and accurate assessment of effects in this key group than was previously possible.

Both analyses used a similar approach. Studies were included if they had used long-term detectors to measure radon concentrations in the living areas and bedrooms of homes currently or previously occupied. Time-weighted average radon exposure was calculated for each participant in terms of becquerels per cubic meter (Bq/m^3) (1 Bq equals one disintegration per second, and 37 Bq/m^3 are equal to 1 pCi/L). Each analysis used a time window for exposure (five to thirty-four years prior to diagnosis of lung cancer in the European analysis and five to thirty years prior to diagnosis in the North American analysis).

In the European analysis, the risk of lung cancer was found to increase by 16 percent (95% confidence interval 5 to 31 percent) per 100 Bq/m^3 in usual radon exposure. This means that those exposed to 100 Bq/m^3 have an odds ratio for lung cancer of 1.16, and those exposed to 200 Bq/m^3 have an odds ratio of 1.32, etc. The large numbers permitted the researchers to assess the effect of radon with fine stratification for smoking history, including average number of cigarettes smoked per day, duration of smoking, and age of starting to smoke. The dose-response appeared linear with no threshold, and there was a significant dose-response relation even below the currently recommended action levels. In the North American analysis, lung cancer risk increased by 11 percent (95% confidence interval 0 to 28 percent) for each increase of 100 Bq/m^3. The odds ratio for the highest exposure level ($\geq$200 Bq/m^3) compared to the lowest level ($<$25 Bq/m^3) was 1.37 (95% confidence interval 0.98 to 1.92).

The emphasis in both analyses is on the finding of an overall association of residential radon exposure with increased lung cancer risk. In both analyses the relationship appears to be linear with no threshold. And both studies note the consistency of the findings with extrapolations from the miner data. There is, however, one striking difference in the presentation of the results of the two papers. In the discussion in the North American analysis, no attention is given to smoking and its interaction with radon. In contrast, the European analysis makes the dramatic and crucial point that, although the increase in the relative risk due to radon is comparable in smokers and never smokers, because the former have so much higher a risk of lung cancer, the absolute risk due to radon is approximately 25 times higher among smokers. This is shown in fig. 5.3 from Darby and colleagues, where in lifelong nonsmokers, the cumulative absolute risk of lung cancer by age 75 due to radon is virtually indistinguishable from zero.

Thus, in line with the BEIR VI report's conclusion based on the miner data, Darby and colleagues' analysis makes clear that the vast majority of excess lung cancer caused by radon occurs in smokers. Even with the greatly increased sample size, it is still unclear what the effects of radon are in never smokers. Curiously, in spite of the evidence from their analysis, Darby and colleagues keep the emphasis on radon, concluding that radon in the home "accounts for about 9% of deaths from lung cancer and about 2% of all deaths from cancer in Europe." And the final sentence of the paper emphasizes the effectiveness of remediation of radon levels in homes, in spite of the questionable impact of remediation on lung cancer rates compared to that of quitting smoking.

At present, efforts are in progress to pool the worldwide studies from North America, Europe, and China.

A DIVERGENT VIEW

Not everyone is comfortable with the reigning consensus regarding residential radon, which could be interpreted as conveying the message that radon at any level is a significant hazard, even in those who have never

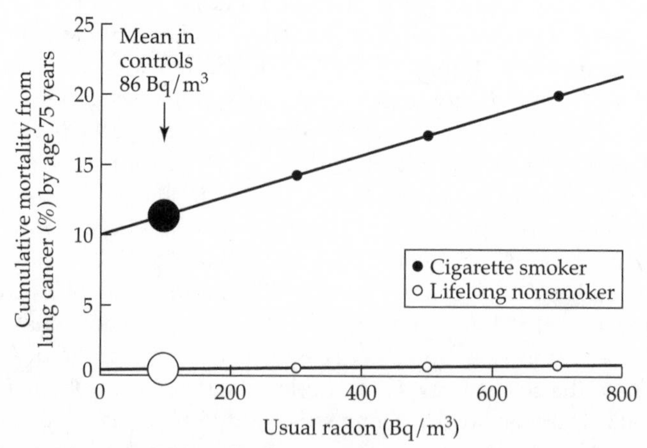

FIGURE 5.3 Cumulative absolute risk of death from lung cancer by age 75 years versus usual radon concentration at home for cigarette smokers and lifelong nonsmokers. *Source*: Darby et al., 2005.

smoked. We have already encountered the views of a number of scientists and observers who voiced strong criticism of what they considered to be a misrepresentation of the effects of indoor radon. Much of this criticism came from health physicists. However, a small number of epidemiologists have also expressed a divergent view and feel that the radon issue has been exaggerated and distorted. These scientists tend to emphasize the considerable difficulties posed by the studies, their methodological limitations, and the overwhelming role of smoking as the dominant cause of lung cancer. They strike a note of skepticism, whereas those who have articulated the consensus position, even while acknowledging the many uncertainties, ultimately imply that that radon, even at very low levels, is an important problem.

But the skeptics are at a distinct disadvantage when it comes to gaining a hearing for their point of view. They tend to be isolated individuals as opposed to the shapers of the consensus, who have the backing of powerful institutions, including the EPA, the National Research Council, and the National Cancer Institute. Several of the scientists standing apart from the reigning consensus pointed out to me how small the number of individuals is that, to a large extent, controls the prevailing view on radon. These individuals, who preside over committees issuing influential reports on radon and who write prominent editorials on the topic, are also likely to review any new paper on the topic for major journals and to review grants in the area of radon. According to several scientists, these figures are unlikely to be sympathetic to researchers who take a more skeptical position. And they are unlikely to find a paper or a grant proposal that takes a critical view of the radon problem to be of sufficient interest to merit publication or funding. As Dale Sandler, who is head of the epidemiology branch of the National Institute of Environmental Health Sciences, remarked, "Conventional wisdom, especially when controlled by a small number of people, is very hard to overcome."[89]

What does the alterative viewpoint look like? For one thing, there is a large area of agreement. There is no dispute that data on underground miners and experimental studies demonstrate that heavy exposure to radon causes lung cancer. The differences have to do with the uncertainty about the effects of exposure at low levels typical of most homes and distinguishing the effects of radon from those of smoking. Ultimately, the dissenters are concerned about putting risks of different magnitudes in perspective and acknowledging the limitations of the existing studies as a basis for risk assessment and policy formulation. Two recent publications articulate this more skeptical and more nuanced point of view.

In 2002, John Neuberger, an epidemiologist at the University of Kansas, and Thomas Gesell, a health physicist at Idaho State University, published a detailed and critical inventory of studies of residential radon exposure, focusing on the results in nonsmokers.[90] Their rationale for concentrating on this group was that "from a public health perspective, the lung cancer risk in smokers can best be reduced through smoking cessation." They noted that smoking rates are declining in the United States and that, if this trend continues, lung cancer rates should decrease in the future, regardless of what is done to control residential radon exposure. In contrast, it is unknown whether radon reduction by itself would provide a significant reduction in risk among smokers. They do acknowledge, however, that nonsmokers exposed to high indoor radon levels could potentially benefit from radon reduction, and this leads to the central concern of their paper—that relatively few studies of residential radon focus on the risk in nonsmokers.

Neuberger and Gesell reviewed the existing studies on residential radon describing their design, summarizing their findings, and identifying limitations. The limitations they discussed include the following: varying definitions used for smoker, ex-smoker, and nonsmoker; failure to account for other potential risk factors; use of surrogate interviews; insufficient sample size; and difficulties of reliably estimating an individual's historical radon exposure. They concluded that a major "challenging problem" even for the better studies is the lack of focus on nonsmokers: "Major public health questions arise for non-smokers, including the effect (if any) of residential radon exposure, the existence of a threshold, and any interaction with passive smoking. The individual residential studies appear to shed little or no light on these issues, possibly because many of them were not designed, nor had the statistical power, to examine these important questions."[91]

The second paper that appears to question the reigning consensus reports the results of a large case-control study carried out by Sandler of the National Institute of Environmental Health Sciences.[92] The study, conducted in Connecticut, Utah, and southern Idaho, was designed in the early 1990s with the aim of improving exposure characterization compared to some of the early studies of residential radon. Specific improvements included obtaining a large sample size, taking into account residential mobility and variations in exposure levels within a home, and imputation of missing measurement data using several sources of data which can help to predict radon levels. A total of 1,474 newly diagnosed cases and 1,811 population controls were included, making this one of the largest studies of residential radon exposure.

In spite of the study's many methodological refinements, the authors found that, "Overall, there was little association between time-weighted average radon exposures 5 to 25 years prior to diagnosis/interview and lung cancer risk."[93] Several of their initial assumptions proved incorrect. For example, actual radon levels in the two states were much lower than had been expected on the basis of surveys. Only 3 percent of homes in Connecticut and 7 percent of homes in Utah/southern Idaho had radon levels exceeding the EPA action level of 4 pCi/L of air. Also, in spite of efforts to maximize the number of nonsmoking lung cancer cases, the authors had too few nonsmoking cases for separate analyses. ("Thus, despite its large size, the study lacked sufficient power to detect main effects at the observed exposure levels and we had limited power to detect interactions.")

In addition to the overall lack of an association between radon exposure and lung cancer, findings within subgroups failed to suggest any meaningful pattern. Sandler and colleagues concluded that, "This study provides no evidence of an increased risk for lung cancer at the exposure levels observed."[94]

Beneath its measured language and careful presentation and discussion of its data, the paper has the earmarks of a corrective to what the authors imply is an overinterpretation of studies of residential radon. This makes it all the more interesting that Sandler and her coauthors note that their findings, although null, are nevertheless consistent with the pooled findings of the seven North American studies. She and three of her coauthors are also authors on that paper.[95] This tension between interpretations of individual studies and of the pooled data underscores the ambiguity of findings from studies of residential radon, which was anticipated by Lubin, Samet, and Weinberg in 1990. Perhaps most provocative is Sandler's mention at the very end of her paper that there may be a level below which radon exposure does not pose a detectable risk: "It is clear from studies of highly exposed populations that radon causes lung cancer. What is less certain is whether or not there is a threshold level below which there is no effect and what proportion of lung cancer can truly be attributed to radon—alone or in combination with cigarette smoking. Only an extremely large study can provide answers to these questions."[96]

Given their assessments of studies of domestic radon, Neuberger and Sandler were less than totally enthusiastic about the undertaking of pooling the various individual studies. Neuberger, who is a co-investigator on the Iowa case-control study of radon, concedes that aggregating the individual studies is a worthwhile exercise but cautions that the results should not be taken as providing definitive answers.[97] He emphasizes that the individual studies were never designed with a view to being combined. They used different methods, different inclusion criteria, different measurement protocols, different quality control, and different definitions of a never smoker and an ex-smoker. For example, Neuberger questions the definition of a never smoker used in the European pooled analysis—someone who has not smoked more than 400 cigarettes per lifetime. He believes that this allows inclusion of some people who smoked at a low level and who should not be classified as never smokers. In addition, he points out that the Darby study is suggestive of a higher risk in men than in women. Because women typically have spent more time in the home and have been less likely to be exposed to occupational carcinogens, this raises a question. Neuberger also notes that Darby's data (fig. 5.3) show a very slight increase in risk for never smokers with increasing radon exposure. However, the graph combines both men and women. Neuberger suspects that the increase may be limited to men, and, for the reasons just mentioned, he feels that women should be analyzed separately. He concludes that the risk estimates that constitute the consensus view on radon "gloss over the fact that the risk of radon-induced lung cancer in never smokers would be minimal or none."[98] Sandler says she is "not enthusiastic about the forthcoming worldwide pooling unless the group can use the combined data to ask more thoughtful questions that deal with how complicated it is likely to be in reality."[99] Thus, the goal of isolating a true effect of indoor radon exposure through the exercise of pooling the worldwide case-control studies seems to be ever-receding.

***

How have the EPA's pubic educational materials changed to reflect the substantial body of scientific evidence on the effects of residential radon exposure that has accumulated over the past fifteen to twenty years? The answer appears to be, hardly at all. Even when the risks associated with indoor radon have been clearly delineated and put in perspective, as was done by the National Research Council in 1999, due to the institutional agenda of the Environmental Protection Agency and the tendency to select numbers that suit its purposes, the key facts surrounding the hazard posed by radon continue to be distorted. The fact that most radon-induced lung cancer deaths will occur in smokers is nowhere explicitly stated even in the most recent EPA literature regarding radon. Instead, the agency resorts to well-intentioned but illogical comparisons, omission of key facts, and impractical proposals to convey the importance of the radon hazard. (I am referring to the recommendation that physicians counsel their patients regarding the importance of testing their homes for radon.) As recently as September 2005, in its *A Citizen's Guide to Radon*, EPA cited the figure of 21,000 lung cancer deaths per year due to radon exposure and displayed a dramatic bar chart comparing deaths from radon to deaths from accidents, including drunk driving, falls in the home, drownings, and home fires.[100]

The arresting graphic (fig. 5.4), which appears on page 2 of the pamphlet, is apparently designed to impress the reader with the magnitude of the radon problem in order to motivate citizens to have their homes tested. However, the bar chart is misleading in two ways. First, it indicates that radon is responsible for more deaths per year than each of four causes of accidental death. But the comparison makes little sense. Unlike accidents, which are instantaneous events, lung cancer due to radon results from exposure over a period of decades. Clearly, it would not have suited EPA's purposes to make the more logical comparison of the number of radon-induced deaths with the much larger number of deaths from chronic dis-

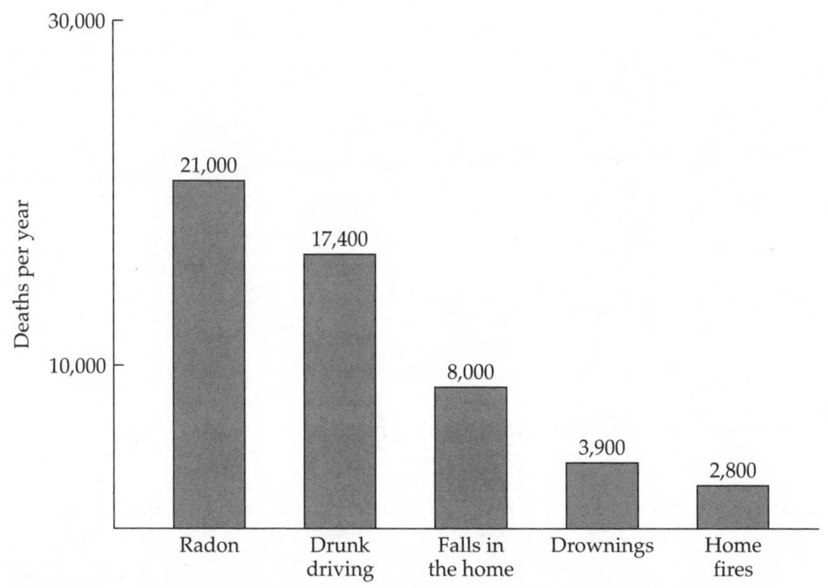

FIGURE 5.4 The U.S. Environmental Protection Agency's estimate of the number of lung cancer deaths in the United States each year due to radon compared to the number of deaths from other causes. *Source*: U.S. EPA, 2005.

eases—say, the total number of deaths from lung cancer, from all cancers, and from coronary heart disease—which would have helped to put this estimate in perspective.

The second aspect of the bar chart, that is even more misleading, is the failure to mention the crucial role of smoking in radon-associated lung cancer. In the brochure's "overview," on page 3, the agency continues to emphasize that radon is the "second leading cause of lung cancer in the United States today," without making any mention of the synergism between radon and smoking. It is only on page 11 that the reader is told that one's "chances of getting lung cancer from radon" depend on "whether you are a smoker or have ever smoked."[101] And page 12 displays a table showing that the risks associated with a given level of radon exposure are roughly seven times higher in smokers compared to never smokers. Even this reference to the "synergism" between smoking and radon, while appropriate, does not really convey to the general reader the crucial fact that roughly 90 percent of radon-induced lung cancers occur in smokers. If the EPA had been concerned to give an accurate picture of the radon hazard, it would have referred to the known lung cancer risk factors and would have distinguished between the effect of radon in smokers and never smokers. The resulting bar chart would have looked like that shown in fig. 5.5, which makes clear that not only is radon-induced lung cancer a

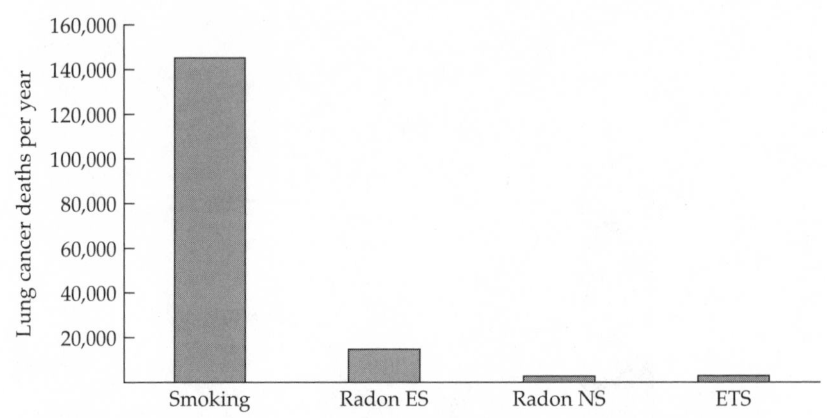

FIGURE 5.5 Estimates of lung cancer deaths from different causes. Estimates of deaths due to radon among ever smokers (ES) and never smokers (NS) come from National Research Council, 1999, p. 15. Estimate of the number of lung cancer deaths due to environmental tobacco smoke (ETS) comes from U.S. EPA, 1992.

distant "second cause of lung cancer" following smoking, but that 90 percent of radon-induced lung cancers occurs in smokers. Estimated radon-induced deaths in never smokers (about 2,000–3,000) represent less than 2 percent of all lung cancers and are comparable to the number of lung cancer deaths ascribed by the EPA to environmental tobacco smoke.

Misinformation regarding radon is not the sole province of the EPA. The confusion surrounding the hazard posed by radon is illustrated by a pie chart purporting to show what is known about the causes of lung cancer in people who have never smoked that was published in a recent issue of the *Journal of the National Cancer Institute.*[102] The graphic in fig. 5.6, which comes from the American Cancer Society, shows that radon and passive smoking each account for a quarter of lung cancer occurring in never smokers. (This estimate for radon is based on BEIR VI.) As we have just seen, there is really no scientific basis for this claim since studies of residential radon are acknowledged to be problematic and since extrapolation from the high levels of exposure in dusty mines in which most miners were smokers to the much lower levels in homes is also problematic. Nevertheless, owing to the tendency to overstate what is known and to want to fill in the pie chart, this "factoid" has received wide circulation. Once such an estimate is put out by an authoritative source, it will be widely cited. In contrast, careful review of the literature on risk factors for lung cancer in never smokers has led several epidemiologists to conclude that we do not know what causes the majority of lung cancers occurring in those who have never smoked.[103] Here is what two prominent epidemiologists have to say in a recent textbook of cancer epidemiology: "Lung cancer among nonsmokers is not rare: it is about as common as cancer of the pancreas.

Causes of Lung Cancer in Nonsmokers

While the overwhelming majority of lung cancer deaths are caused by smoking, thousands of people who never smoked die of lung cancer. The American Cancer Society estimates that about 11,000 never-smokers in the United States die of lung cancer annually. Research on the nonsmoking causes of lung cancer has focused primarily on second-hand smoke, radon, asbestos and other occupational causes, outdoor air pollution, and genetics. Estimates of the deaths attributable to these causes, independent of smoking, is controversial in some cases, such as asbestos.

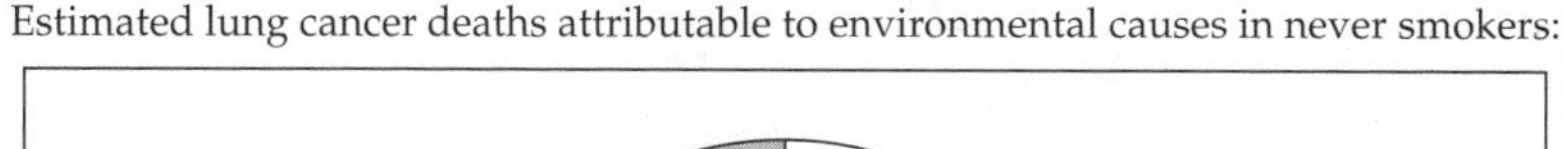

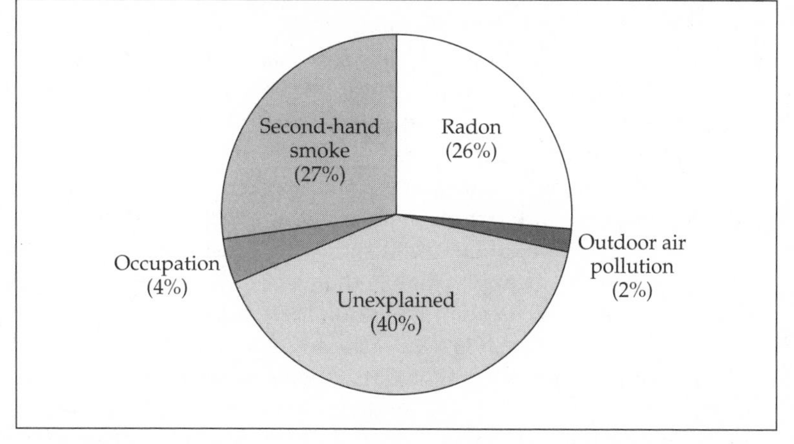

FIGURE 5.6 Causes of lung cancer in nonsmokers. *Source*: Stat Bite 2006:664.

Occupational factors, passive smoking, and indoor exposure due to radon do not explain more than a minor proportion of these cases, and dietary, infectious, and genetic factors are currently receiving attention."[104] And in a comprehensive review of the causes of lung cancer in nonsmokers, a group of researchers actively involved in studies of radon and lung cancer arrive at a similar conclusion: "it is clear that the largest contributors to the etiology of lung cancer in nonsmokers have not been elucidated."[105]

***

Efforts to pin down the effects of residential radon are reminiscent of the movie *Blow-up* by Michelangelo Antonioni in which a photographer continually enlarges a photograph in which he may have unwittingly recorded a murder. The more he enlarges the photo, the grainier it gets, and the evidence he seeks continues to elude him. This is an apt metaphor for efforts to document effects of indoor radon exposure at what are, on average, very low levels. (The average level in U.S. homes is only slightly more than twice the concentration in outdoor air.) It should be possible to acknowledge that radon at such levels could theoretically pose some additional risk of lung cancer but that all available evidence indicates that this increment of risk is going to be small. According to the radiation expert John Boice, the average relative risk of lung cancer due to exposure to environmental radon is on the order of 1.2.[106] (This includes smokers, as well as never smokers). This relative risk, which is equivalent to a 20 percent increase in risk, needs to be put in perspective by comparing it to the risk of lung cancer due to smoking—about 20-fold for a male current smoker and about tenfold for a male ex-smoker, or a 2,000 and 1,000 percent increase, respectively. The lung cancer risk of a heavy smoker (of more than two packs per day) can reach 4,000 or 6,000 percent. Instead, we have seen how strong is the tendency to make radon into a more substantial problem than it likely is. Scientists and regulators with a professional stake in radon may not be in a position to give an objective assessment of its importance. It should be possible to acknowledge that much is known about the effects of radon at high levels encountered in mines but that we cannot characterize the risk at the very much lower levels typical of homes. It should be possible to acknowledge uncertainty where this is appropriate, where everything tells us that this is an area of uncertainty, beyond the resolving power of currently available methods and data. But it is a lot easier to come up with estimates of a number of deaths ("the body count") attributable to an agent, even if these are questionable. This satisfies a deep desire in the regulator and scientist. It provides a justification for their efforts and for large amounts of money spent on radon abatement and research. After all of the attention devoted to radon, few are going to be satisfied with a conclusion that says, "We don't know what the effects are at these low levels. There may be a very small effect, or possibly there may be no effect."

Box 5.1 Major Take-Home Points

Radon is a radioactive gas resulting from decay of radium and ultimately from uranium that occurs naturally in rock and soil. Radon can seep into homes through cracks in the foundation and, under certain conditions, accumulate to reach high concentrations. Radon decays to form "progeny," which emit a type of radiation capable of being inhaled and causing cancer in the cells lining the lung.

It has been known since the early twentieth century that underground miners exposed to high levels of radon gas have an increased risk of lung cancer. However, the possibility that residential radon could pose a hazard only became a subject of widespread concern in the mid-1980s, when extraordinarily high levels were detected in a home.

In spite of the paucity of data on the effects of radon exposure at the generally low levels found in homes, in 1986 the U.S. Environmental Protection Agency set an "action level" of 4 pCi/L for single-family homes, advising that levels exceeding 4 pCi should be reduced. On the basis of early, nonsystematic surveys of radon levels, 8 million homes were estimated to be in need of remediation.

Radon was labeled "the second leading cause of lung cancer" and was estimated to cause between 6,000 and 30,000 lung cancer deaths per year in the United States. However, it was rarely made clear that roughly 90 percent of deaths attributed to radon are likely to occur in smokers and would not have occurred in the absence of smoking.

Because studies of underground miners constituted the major source of information on human exposure to radon, elaborate analyses were carried out to extrapolate downward from the effects of exposure in miners to those in the general population due to residential exposure. However, this type of extrapolation entails a number of assumptions that may not be valid. Most importantly, miners had much higher exposures and the majority of miners were smokers. Furthermore, the conditions in underground mines were very different from those in homes.

Since the early 1990s, studies of residential radon exposure have been carried out. Such studies are subject to serious problems of measuring radon exposure over a period of decades as well as detecting an effect of radon independent of the effect of smoking. The individual studies show inconsistent results, and pooling of the individual studies has been undertaken to provide a more stable estimate. Two major poolings indicate a small increase in risk associated with increased radon exposure. However, again, roughly 90 percent of the effect of radon is accounted for by smokers, and effects in never smokers remain unclear.

Efforts to assess the impact of residential radon on lung cancer continue. However, there is tendency to avoid putting the risk due to domestic radon in perspective, and, consequently, the hazard, particularly to never smokers, appears to have been greatly overblown.

6

THE CONTROVERSY OVER PASSIVE SMOKING
A Casualty of the "Tobacco Wars"

As far as I know, there's no legitimate scientist in the world who doesn't think secondhand smoke causes lung cancer and heart disease.

—Stanton Glantz, 2003

The main way smokers kill people is by smoking themselves, not by killing other people—they are a lot better at killing themselves than they are at killing other people.

—Sir Richard Peto, 2006

In the late 1990s a billboard along California highways showed a handsome young man with an unlit cigarette in his mouth facing an attractive young woman. The man is asking the woman: "Mind if I smoke?" And the woman's carefully counterpoised rejoinder is: "Care if I die?" The notion that exposure—even the most casual exposure—to secondhand tobacco smoke could be lethal to a nonsmoker had achieved such currency that the California Department of Health Services could evoke it very effectively with this minimal but pointed social exchange. California's eye-catching billboard is just one indicator of the attention and resources that have been devoted to the issue of the health effects of exposure to secondhand tobacco smoke. But how does the billboard's message and the perception of the lethality of exposure to passive smoking square with the scientific evidence? How did environmental tobacco smoke become the high-profile cause that it is today and what are the deeper determinants behind the campaign for a smoke-free environment? Finally, how is the interested observer to make sense of what is behind the steady succession of studies and reports and claims of serious health effects?

As I will make clear, the attention and resources devoted to the question of environmental tobacco smoke cannot be explained solely by reference to the science. Observers as different as Richard Kluger, Michael Crichton, and Peter L. Bernstein[1]—none of whom is known to be a lackey of the tobacco industry—have pointed out the weakness of the case for an association of exposure to secondhand smoke and lung cancer in nonsmokers. Some prominent scientists have also registered their skepticism that exposure to environmental tobacco smoke (ETS) is a cause of serious disease.[2] But, for reasons that I will explain, these independent voices simply could not compete with the massive campaign mounted by the antismoking movement and its allies in government and academia. Each milestone in the campaign against ETS has led to further claims, often more tenuous than the preceding ones. The mountain of reports and scientific literature enshrining this dogma is so formidable that it is hard to see how anyone could question it. Furthermore, the terms of the debate have been carefully orchestrated so that any criticism or attempt to reconsider the scientific basis for the claims is invalidated from the start.

I should make clear at the outset that my account of the ETS story has nothing to do with questioning whether exposure to secondhand tobacco smoke could cause some additional cases of lung cancer and other diseases in people who never smoked. ETS contains many of the same carcinogens and toxins present in mainstream tobacco smoke, and it is entirely biologically plausible that nonsmokers who are chronically exposed to high levels of ETS and those who have a constitutional susceptibility are at increased risk. In addition, secondhand tobacco smoke is associated with increased incidence of lower respiratory infections in infants and children, exacerbation of preexisting asthma, and other effects. Nor do I have any objection to the most stringent restrictions on smoking—to my mind they are entirely justified and should have been enacted long ago. The only question at issue here is whether on a topic as important and as sensitive as ETS all of the science should be allowed to be heard, or whether it is acceptable to select and slant the science in order to justify a particular policy. It is a question of the means that are used to achieve a beneficial end.

Because the data are weak and are not all that scientists and regulators would like them to be, one of the characteristics of "expert opinion" and discussion on this topic in general has been to not look too carefully at the data. This deliberate willingness to be careless with data—and, even worse, the willingness to go on to the next step by distorting and misrepresenting the findings of scientific studies—is the mainspring of a political strategy of hyping an unrealistic risk in order to gain public support for tobacco control policies. As such policies become ever more extreme, they in turn, by an irresistible feedback mechanism, require ever more extreme data, or interpretations of the data, to support them.

Given this background, it is not surprising that little attention has been devoted to the way in which weak epidemiological findings have been overstated in the scientific literature and in government and international agency reports. And yet, such an analysis has much to tell us about the sociology of public health science and, specifically, how science can be misused for what are political ends—worthy though they may be. By recalling the origins of the campaign against ETS and particularly by critically examining key documents and arguments designed to influence public policy on this issue over a period of twenty-five years, we can gain an understanding of how what has been presented as "science" has been shaped by extra-scientific agendas.

HARM TO THE BYSTANDER

At the outset a few preliminary remarks are in order. First, secondhand tobacco smoke, or environmental tobacco smoke, is composed of side-stream smoke from the burning tip of the cigarette plus the smoke exhaled by the smoker. Over time this mixture ages and is deposited on surfaces. Second, the effects of exposure to secondhand smoke can only be studied in people who have never smoked since any effects would be dwarfed by those of active smoking, which involves the inhalation of smoke directly into the lungs where it is much more concentrated. It is worth noting, however, that active smokers would be expected to have the greatest exposure to ETS. If we truly believe in the concept of a dose-response (i.e., the dose makes the poison, to loosely paraphrase Paracelsus, and as countless epidemiological studies of active smokers have confirmed), then the dosage received by a passive smoker must be a miniscule fraction of that received by an active smoker, and, therefore, it is preposterous to think that passive smokers have a risk that comes within an order of magnitude of an active smoker.

The possibility that exposure to secondhand tobacco smoke might cause adverse health effects was first articulated in the early 1970s, well before virtually any scientific evidence was available on this question.[3] In 1971, Surgeon General Jesse L. Steinfeld cited accumulating "evidence that the nonsmoker may have untoward effects from the pollution his smoking neighbor forces upon him," and he called for a ban on smoking in "all confined public spaces such as restaurants, theaters, airplanes, trains and buses." The following year, the *Surgeon General's Report on Smoking and Health* for the first time labeled nonsmokers' exposure to cigarette smoke as a health hazard.

Even before the Surgeon General's report, surveys had already shown that 58 percent of men and 72 percent of women who had never smoked believed that smoking should be reduced in public places, and more than three quarters of never smokers considered it "annoying" to be in proximity to smoking. Responding to the increasing awareness of tobacco smoke as a nuisance and possible health hazard, the nonsmokers' rights movement emerged in the 1970s. One of its leading organizations was GASP (Group Against Smokers' Pollution), which effectively lobbied for restrictions on smoking in public. The movement's early successes testified to changing attitudes toward smoking and the effectiveness of the appeal to nonsmokers' right to breathe unpolluted air. Early achievements included the provision of separate seating for smokers and nonsmokers on domestic airliners (1973), the restriction of smoking on interstate buses (1974), and restrictions on smoking in public places by states and localities, starting with Arizona, Connecticut, Minnesota, and Berkeley, California.

It did not take long for the tobacco industry to realize that the growing effectiveness of the nonsmokers' rights movement represented a fundamental threat to its existence. As articulated in a 1978 report by the Roper Organization undertaken for the industry, almost 60 percent of survey respondents agreed that smoking was probably harmful to nonsmokers. Furthermore, a surprising 40 percent of smokers believed that their smoking was a hazard to nonsmokers. The authors of the report struck an alarming note: "We believe it would be difficult to overemphasize the importance of [these] finding[s], indicating as [they do] that the battle to convince the public of the dangers of passive smoking is in the process of being lost, if indeed it is not already over."[4]

Thus, it is important to note that, from the beginning, the issue of secondhand smoke was inextricably intertwined with both the campaign for clean air and the struggle to reduce smoking and smoking-related disease. This has meant that scientific research to elucidate the effects of exposure to cigarette smoke on nonsmokers—which began to appear only after the movement was well under way in the 1970s—has been subject to enormous pressures to provide supportive data. As a result, the evidence from the studies on this question has often been overstated, taken out of context, and interpreted selectively by the opposing forces on both sides of the "tobacco wars" to suit their purposes. Although it was almost never forthrightly acknowledged by scientists focusing on passive smoking, the stakes hinging on the interpretation of exquisitely small excess risks in the epidemiologic studies were enormous.

Rather than overstating the science and asserting dogmatically that the risk from ETS is established, it would have been more appropriate and less misleading to state that it is entirely plausible that in some cases exposure to ETS may account for a few cases of lung cancer in nonsmokers, but that, owing to the problems that will be discussed below, it is not possible to quantify the excess risk with certainty. It is the false impression of certainty and of precision that is most objectionable in the series of agency reports and meta-analyses on secondhand smoke.

The pattern of omissions and distortions of the official consensus becomes intelligible when one realizes that the goal was not to clarify the issue and provide an honest assessment. A careful reading of the 1992 EPA report and a number of subsequent documents on the health effects of secondhand smoke makes it abundantly clear that the passive smoking issue is being used for an ulterior purpose—namely as an indispensable weapon in the campaign to reduce the prevalence of smoking. If secondhand smoke can be shown to be a hazard to nonsmokers, increasing restrictions on smoking in public places will severely limit the opportunities for smokers to smoke. Smokers will, as one tobacco industry memo put it, increasingly "have nowhere to hide." Thus, the real agenda went beyond the narrow scientific one of critically evaluating the available evidence on ETS. The real goal was to stamp out smoking. This is the crux of the passive smoking issue and explains much of the confusion surrounding the issue.

Starting with the 1986 reports and especially with the 1992 EPA report, *suggestive evidence of a possible slight increase* in the risk of lung cancer was used to give teeth to legislation restricting smoking in public places. The fact that secondhand smoke is an irritant and an annoyance, that it is associated with increased respiratory infections in infants, and that it exacerbates pre-existing asthma and other health conditions simply does not provide the same legal or regulatory clout as the claim that it causes fatal disease. This explains why it has been hard for scientific findings regarding secondhand smoke to be interpreted in a disinterested manner. To acknowledge that the data are weak—as they would have to be, given the low concentration of ETS and the limitations of observational studies on this question—has been anathema because this would deprive the antismoking movement of its most powerful weapon against the tobacco industry. The tactic of presenting massive amounts of data devoid of any critical framework for making sense of those data was meant to obscure this sleight of hand. In large part, scientists and regulators have relied on categorical pronouncements and on the inherent obscurity of the material to create an unassailable dogma. Who could possibly question the wisdom of such authorities as the U.S. Surgeon General, the Environmental Protection Agency, and the World Health Organization?

ETS AND LUNG CANCER

Since 1981, when the first epidemiological study to address the question of ETS and lung cancer risk was published, at least 76 individual studies have been published worldwide.[5] In addition, at least 20 meta-analyses have been carried out and numerous reports have been issued by government agencies and international health agencies. Rather than discuss this whole body of literature, my goal here is first to describe the early studies, which set the stage for epidemiologic research on this question, and then to identify some of the problems with the epidemiologic studies that have received scant attention in the scientific literature and in the consensus reports. With the exception of the first study, which was carried out in Japan, I focus on U.S. studies because these are most germane to setting policy in this country. Furthermore, special attention is devoted to the American Cancer Society prospective studies, which are important because of their size but also because these studies play a central role in a controversy concerning the association of ETS and heart disease, which will be discussed later in this chapter.

In order to understand what was made of this body of evidence, it is necessary to have a sense of the quality of the information that was used to assess ETS exposure. Virtually all studies of a link between exposure to secondhand tobacco smoke and lung cancer or other serious disease have relied primarily on marriage to a smoker as a surrogate for regular exposure. (In addition, some studies collected information about exposure to

ETS at work and in other settings.) Thus, one typically focuses attention on people who have never smoked and contrasts the risk of disease among those whose spouses smoked with that of those whose spouses were never smokers. While it is reasonable to assume that nonsmokers married to smokers have greater exposure than nonsmokers married to never smokers, it is important to recognize how distant this marker is from providing an accurate measurement of how much smoke a given individual was exposed to over his or her lifetime, which is what one would like to know. Even if two different individuals lived with spouses who smoked the same number of cigarettes per day, they could have very different exposures depending on the number of hours spent in the home, where each spouse spent time within the home, the size of the rooms in which smoking took place, ventilation, and so on. Furthermore, people could have exposure from other family members they lived with and in other settings, including work, social settings, and in transportation, and exposure in these venues is also difficult to assess by questionnaire.

Case-control studies of passive smoking and lung cancer in nonsmokers—and the majority of studies on ETS and lung cancer are case-control studies—are prone to a particularly serious form of bias. A nonsmoker who is diagnosed with lung cancer—a disease so strongly linked to active smoking—is likely to want to identify what could have caused his or her illness. When an interviewer inquires about exposure to other people's smoking in the home, at work, and in social settings, the nonsmoker diagnosed with lung cancer may have an inherent tendency to recall more diligently and more thoroughly past exposures—or to overestimate their intensity or duration—compared to a healthy control, or a hospitalized control with a diagnosis unrelated to smoking. In other words, the case may tend to "ruminate" about past exposures to a greater extent than the control, because the control does not have the same motivation to come up with an explanation. Thus, information on exposure to ETS obtained in case-control studies is particularly susceptible to "recall bias." It does not require very extensive overreporting of exposure to passive smoking on the part of the cases to produce a small but spurious association. Awareness of the possibility of recall bias in case-control studies of ETS is a fundamental aspect of interpreting epidemiological studies on this topic, but it was frequently given short shrift in the rush to interpret these studies as providing evidence of a causal association.

One important strength of cohort studies on this topic is that they are not affected by recall bias, since information on smoking habits is obtained from individuals who are healthy at entry into the study and who are then followed for a number of years. Furthermore, in cohort studies, information on smoking habits is obtained from each member of a married couple, so that ETS exposure is based on the spouse's smoking habit. However, cohort studies on this question also have drawbacks, including the fact that in most instances information on exposure prior to enrollment was not obtained. In addition, information on exposure from sources other than the spouse, information about previous marriages, and information on changes in exposure following enrollment is generally lacking. This will likely result in misclassification of exposure.

Early Studies

The first epidemiological study to suggest that exposure to secondhand smoke might pose a serious threat to health was the now famous study from Japan entitled "Non-smoking wives of heavy smokers have a higher risk of lung cancer," which was published in the *British Medical Journal* in early 1981.[6] The author Takeshi Hirayama made use of data from a cohort study of 91,540 nonsmoking wives whose husbands' smoking habits were known. The women in the study were followed for fourteen years, and the mortality rate among women whose husbands were smokers was compared with that of women whose husbands were nonsmokers. Women whose husbands were either ex-smokers or current smokers of 1–19 cigarettes per day had a relative risk of 1.61, and women whose husbands were current smokers of 20 or more cigarettes per day had a relative risk of 2.08, or an apparent doubling of their risk of lung cancer, compared to women whose husbands were nonsmokers.

Publication of the Hirayama study received front page coverage in newspapers and raised the question of the effects of exposure to passive smoking as a cause of fatal disease for further study by epidemiologists, analytical chemists, and respiratory toxicologists. The paper also elicited a number of letters to the journal, many of them critical of Hirayama's presentation and his methods. Undoubtedly, the paper's brevity—it was only two-and-a-half pages long—accounted for some of the difficulty readers had in assessing the quality of the data and the analysis. But by far the main question pertaining to this study and all subsequent studies of the effects of exposure to passive smoking on serious disease is that of assessing actual exposure. As mentioned earlier, a husband's smoking habit provides only a poor surrogate marker for the actual exposure of the nonsmoking spouse, that is, for how much smoke was actually inhaled and deposited in her lungs. Use of this surrogate rests on the assumptions that (1) if the husband is a smoker the spouse's exposure will be greater than if the husband is a nonsmoker and (2) the more a husband smokes, the greater his wife's exposure will be. However, these assumptions may not be valid. First, the nonsmoking spouse may have been exposed to other important sources of secondhand smoke. Second, in Japan, there is a long-standing custom of men socializing with their male coworkers outside the home after work. Thus, it is not clear how much time the typical wife was in the presence of her smoking husband. In the Hirayama study and in most subsequent studies, no information is available on how much time the husband and wife were typically in each other's presence, how much smoking went on during this time, or on the conditions under which exposure took place (size of the room, ventilation, etc.). Another question regarding the Hirayama study is how the husband's smoking habits may have changed over the fourteen-year follow-up period after baseline questionnaire data were obtained.

Taking his results at face value, Hirayama estimated that the effect of passive smoking was roughly one third to one half that of active smoking on lung cancer risk, even though this would appear implausible, given the great difference in the concentration of smoke inhaled by the nonsmoker and that inhaled by the active smoker. He also proposed that his results helped explain at least in part the "long-standing riddle" that lung cancer rates among Japanese women had been increasing in the post-war period in spite of the fact that only "a fraction of Japanese women with lung cancer smoke cigarettes." Hirayama's inclination to interpret the results of the first study on passive smoking and lung cancer as indicative of a causal association, in spite of the many limitations of the data and study design, is noteworthy in view of the way in which subsequent studies on this topic have been interpreted and used.

Only a few months following publication of the Hirayama paper, Lawrence Garfinkel published an article in the *Journal of the National Cancer Institute* entitled "Time trends in lung cancer mortality among nonsmokers and a note on passive smoking."[7] Garfinkel was the head of epidemiology at the American Cancer Society (ACS) and had designed its first prospective study initiated in 1959 together with E. Cuyler Hammond. As noted earlier, the idea that exposure to secondhand tobacco smoke could be harmful to the nonsmoking bystander had been "in the air" since at least the mid-1980s. Part of Garfinkel's motivation for examining time trends in cancer mortality among nonsmokers came from a paper by a University of California, Los Angeles (UCLA) epidemiologist, James Enstrom, who two years previously had published a paper examining lung cancer rates in nonsmokers in the United States from 1914 to 1975. Enstrom had concluded that the apparent increase in lung cancer mortality in male nonsmokers suggested that factors in addition to personal cigarette smoking contribute to the disease and that "a more complete understanding of lung cancer etiology is needed."[8]

Garfinkel calculated lung cancer mortality rates among nonsmokers in two large prospective studies: the American Cancer Society's own study and the Dorn study of U.S. veterans. The ACS study contained 94,000 male and 375,000 female nonsmokers at the start of the study. (A "nonsmoker" was one who reported that he or she had never smoked or smoked only occasionally but had never smoked regularly.) Study participants filled out four additional questionnaires after the start of the study, and very few nonsmokers reported that they started to smoke on any of the four later questionnaires. About 54,000 of the veterans in the Dorn study were nonsmokers (the same definition of a nonsmoker was used as in the ACS study).

Garfinkel's analysis showed no change in the lung cancer death rate among nonsmokers in either males or females in the ACS study over a twelve-year period or in the male veterans over a fifteen-year period. The results for women were of particular interest, since historically men tended to be heavier smokers and had started smoking earlier in the century, and, therefore, nonsmoking women married to smoking husbands would be expected to have greater ETS exposure than nonsmoking men married to smoking wives. The lung cancer death rates (per 100,000 "person-years") among nonsmoking women in the ACS cohort for three time segments between 1960 and 1972 were 13.8, 12.9, and 13.1, showing no indication of an increase.

In addition to examining lung cancer death rates among nonsmokers, Garfinkel carried out an analysis similar to Hirayama's, comparing lung cancer deaths among nonsmoking women whose husbands were cigarette smokers to deaths among women whose husbands were nonsmokers. Like Hirayama, he reported the relative risk for two exposure levels: for women whose husbands smoked less than twenty cigarettes per day and for women whose husbands smoked more than twenty cigarettes per day—both compared to women whose husbands were nonsmokers as the reference group. (These results appropriately took into account a number of sociodemographic factors.) The relative risk for the lower exposure group (husband smoked fewer than twenty cigarettes per day) compared to the unexposed was 1.37, while that for the higher exposure group (husband smoked twenty or more cigarettes per day) was 1.04. Neither risk estimate was statistically significant. What is most striking is that there was no suggestion of increased risk in the group whose husbands were heavy smokers.

In his discussion, Garfinkel referred to the central problem of the early (and later) studies of passive smoking, i.e., that of "misclassification of exposure status." In other words, women whose husbands smoked might have minimal exposure to their husband's cigarette smoke, whereas they may have other significant sources of exposure, and conversely, women whose husbands are nonsmokers may be exposed from other sources. Such misclassification, if it is random, would lead to an obscuring of an association due to passive smoking. Owing to this problem, he cautioned that the results of his analysis should be interpreted cautiously.

But Garfinkel also proceeded to refer to the results of autopsy studies carried out in cigarette smokers and nonsmokers.[9] In cigarette smokers there was a "dose-related spectrum" of abnormal cellular changes in the lung that are precursors of lung cancer. In contrast, serious changes in the lung were found in less than 0.1 percent of the specimens of nonsmokers. This appeared to cast doubt on the possibility that exposure to other people's cigarette smoke could be an important cause of lung cancer in nonsmokers. ("Since there is such little variation in the appearance of these histologic changes in nonsmokers of different age, sex, and residence, it seems doubtful that those nonsmokers who had been heavily exposed to cigarette smoke from others in their lives could have had many more precursor lesions for the development of lung cancer than nonsmokers not so exposed."[10]) And Garfinkel concluded that, "Therefore, there is evidence from these studies that passive smoking cannot play more than a very small role in the development of lung cancer."[11] In reference to the Hirayama paper, Garfinkel wrote that, "It appears unlikely on a biologic basis, therefore, that wives with husbands who smoke 20 or more cigarettes a day can have mortality ratios that approach those of regular cigarette smokers."[12] And in a 1984 paper on "Passive smoking and cancer: American experience," he came right out and stated that, "The American Cancer Society's prospective study have failed to show such a risk [i.e. ETS-related to lung cancer]." [13]

Garfinkel's 1981 assessment of the question of whether exposure to passive smoking causes lung cancer has the hallmarks of a dispassionate, scientific approach. He brought to bear on the question data from different sources—lung cancer time trends, autopsy results, and epidemiological findings, weighing what each type of evidence had to contribute. He considered the biases that were likely to affect epidemiological studies on this question. And he invoked biological plausibility—that is, it is difficult to see how the effects of ETS exposure could approach those of active smoking, as reported in Hirayama's results. He noted that the studies published up to that time had not been designed to answer the question of the effects of ETS exposure and emphasized the need for studies with improved methodology to assess the issue. What is striking is the independence of mind exhibited in this early work and the willingness to entertain any pertinent evidence to get a fix on the question. It is clear from Garfinkel's approach that the scientific question concerning ETS needs to be resolved using the best data obtainable and that this question—is there a detectable effect?—is logically prior to and separate from considerations of public health. Rereading Garfinkel's early papers today brings home to what extent the scientific question was later to become muddied by evangelical public health and antismoking campaigns.

The Hirayama and Garfinkel papers from 1981 represent two very different views on passive smoking—the former interpreting the result as reflecting a causal association; the latter approaching the issue with skepticism and bringing to bear a variety of different types of evidence.

Early Agency Reports

Nineteen eighty-six saw the publication of the first three government and international health agency reports to address the health effects of passive smoking. These were from the U.S. Surgeon General, the National Research Council (NRC), and the International Agency for Research on Cancer (IARC), an arm of the World Health Organization.[14] All three reports reviewed the published studies on the topic of ETS and lung cancer, but there are differences in emphasis. Both the NRC report and the Surgeon General's reports concluded that exposure to ETS is a cause of lung cancer in nonsmokers. The NRC report included a meta-analysis of the thirteen studies of ETS published as of that time that yielded a summary relative risk of lung cancer of 1.3 for nonsmokers exposed to ETS compared to unexposed nonsmokers. Only a brief section of the 1986 IARC report on tobacco smoking was devoted to ETS, and IARC was more restrained, acknowledging that an effect of ETS on lung cancer was biologically plausible but difficult to detect owing to the much lower concentration of ETS compared to the active smoker's exposure.[15] The 1986 Surgeon General's report drew the strongest conclusion of the three reports and provided a preview of the argument used in the 1992 EPA report. A key paragraph of its "overview" states that:

> Cigarette smoke is well established as a human carcinogen. The chemical composition of ETS is qualitatively similar to mainstream smoke and sidestream smoke and also acts as a carcinogen in bioassay systems. For many nonsmokers, the quantitative exposure to ETS is large enough to expect an increased risk of lung cancer to occur, and epidemiologic studies have demonstrated an increased lung cancer risk with involuntary smoking. In examining a low-dose exposure to a known carcinogen, it is rare to have such an abundance of evidence on which to make a judgment, and given this abundance of evidence, a clear judgment can now be made: exposure to ETS is a cause of lung cancer.[16]

None of the three reports devoted much attention to the problem of assessing an individual's lifetime exposure to ETS, and they tended to take the results of the epidemiologic studies at face value. There was little critical concern for differences in exposure in countries with very different living conditions or for the possibility of reporting bias in case-control studies or for the problem of confounding, which, as we have become acutely aware in the past few years, can seriously affect the results of observational studies in general. Reading these reports, one cannot escape the impression that the overall willingness to interpret weak data as evidence of causality would not have passed muster if the topic had not been tobacco.

The 1992 EPA Report

Publication of the 1992 EPA report entitled *Respiratory Health Effects of Passive Smoking*[17] represents a landmark in the campaign against ETS. This document was to have a much greater impact, as it was designed to have, owing to its high visibility and the fact that its conclusions were accompanied by actual estimates of the number of deaths and cases of disease ascribable to exposure to secondhand tobacco smoke. The report's major conclusion stated that ETS was a "known human carcinogen" and was responsible for approximately 3,000 lung cancer deaths annually in adult nonsmokers in the United States.

Although the EPA had no mandate to regulate exposure to cigarette smoke, the agency was aware that the existence of a strong document linking exposure to secondhand smoke to disease would provide a powerful tool for other agencies, like the Occupational Safety and Health Administration (OSHA), and local jurisdictions that did have the authority to enact restrictions on smoking. The EPA was also aware that a report given wide publicity would provide a critical touchstone—a founding text—for those wanting to cite an authoritative source in support of the claim that exposure to secondhand smoke caused serious health effects.

The EPA report represents an enormous amount of earnest and useful analysis of the available data as of that time. Some of the epidemiologic evidence presented is indeed suggestive of a weak association of ETS with lung cancer. My criticism of the report boils down to the fact that the agency should have been more forthright about acknowledging some of the weaknesses and inconsistencies in the data and also in citing evidence that did not provide support for an association of ETS with lung cancer. As it is, the presentation is slanted in order to provide support for a specific policy.

It is significant that the EPA viewed its task as being to "update" the 1986 National Academy of Sciences and Surgeon General's reports on passive smoking by marshalling the new studies that had appeared since those reports were published. In other words, the EPA knew what the objective was and viewed the task of drafting what it intended to be an authoritative report more as a mechanical task than as a rigorous and impartial review of all of the scientific evidence relevant to the question at hand. In fact, it is clear that from the outset in the late 1980s, the EPA had decided that the report would classify ETS as a *known human, or Class A, carcinogen.*[18] In a sense, everything else was window dressing.

Echoing the 1986 Surgeon General's report, a frequent refrain throughout the EPA document is that it is rare to have such a wealth of data available on "actual" real-world exposure to ambient levels of a toxic agent as in the instance of environmental tobacco smoke.[19] This often repeated comment refers to the fact that, in setting limits for exposure to an environmental pollutant for the general population, it is often necessary to make use of data from occupational studies, which provide information on the effects of high levels of exposure. One then extrapolates downward to estimate the effects of exposure of the general population to considerably lower levels. However, what this refrain, repeated like a mantra, reveals is the EPA's tendency to minimize the real difficulties of assessing an individual's lifetime exposure to something as diffuse and variable over time as secondhand tobacco smoke. The authors of the report preferred to avoid this issue.

Some of the major omissions and questionable procedures in the EPA analysis concerning ETS and lung cancer are listed in box 6.1.

Points like those raised in box 6.1 are relevant considerations that should have been included in the EPA report to convey a complete and honest picture of this potential hazard.[21] And once included in the report they should have been taken into account in arriving at a conclusion.[22] The failure to acknowledge such considerations is really symptomatic of a more fundamental failure to put the issue of passive smoking in perspective. If

Box 6.1 Some Major Omissions and Deficiencies of the 1992 EPA Report

It should be emphasized that lung cancer in never smokers is an extremely rare occurrence—on the order of 10–15 cases per hundred thousand. If passive smoking carries a 20 percent increase in risk, this would mean an additional 2–3 cases of lung cancer in never smokers per 100,000 population.

When dealing with risks of such a low order, we are at the limits of what epidemiologic studies can reliably detect.

For perspective, the 20 percent increase in risk associated with passive smoking needs to be compared with the 2,000 percent increase in risk for current smokers compared to never smokers and with the 4,000–6,000 percent increased risk in heavy smokers.

Lung cancer rates in never smokers remained essentially flat at roughly 10–13 cases per 100,000 between 1959 and 1982.[20]

Need to acknowledge the difficulty of assessing an individual's exposure to ETS over a lifetime, because exposure varies over time and in different situations.

Average dose of ETS to which the nonsmoker is exposed is much lower—likely by 3 orders of magnitude—than that of the average active smoker.

Both case-control and cohort studies have serious limitations in measuring ETS exposure.

Pathologic changes in the lung tissue of male smokers and nonsmokers were compared in men who died of causes other than lung cancer in 1955–1960 and 1970–1977. Smokers had high rates of precancerous changes, and the frequency of these changes increased with the amount smoked. In contrast, only 0.1 percent of nonsmokers had precancerous changes, suggesting that ETS is not an important cause of lung cancer in never smokers.

The agency used what appear to be a low rate of "smoker misclassification" and a greatly inflated factor to adjust the relative risk due to spousal smoking for background ETS exposure (see p. 165, this volume).

Considering these and other weaknesses, the evidence concerning ETS may not justify the classification of Class A, or "known human carcinogen" in the same category as tobacco smoke (active smoking), benzene, arsenic, radon, etc.

the EPA report had been first and foremost about science, the issue of ETS could easily have been put in perspective by comparing the association of ETS exposure with lung cancer in never smokers with the lung cancer risk associated with active smoking. In addition, the strength and coherence of the evidence in the two cases would have been compared. This most basic step was studiously avoided by the redactors of the document. One can only conclude that the EPA report and most subsequent reports have found it essential to blur the crucial distinction between active and passive smoking. To give it its due might weaken the case against ETS.

Another crucial distinction that the agency found it convenient to blur was that between association and causation. As is well understood by epidemiologists, the fact that studies generally found a slight increase in the risk of lung cancer associated with exposure to ETS is not proof of a causal association. Because of the small magnitude of the excess (20–30 percent), it is possible that some, or all, of this excess could be explained by various biases or confounding. When the factor under study is as difficult to measure as ETS, an association represented by a 25 percent increase in risk of lung cancer in nonsmokers is very likely at the limit of what can reliably be sorted out in observational epidemiologic studies. This would be routinely acknowledged by epidemiologists, if the topic under discussion were any other than passive smoking. These two intentional obfuscations on the part of the EPA are at the core of what some commentators have referred to as the agency's "betrayal of science."[23]

Essential to the campaign against ETS all along has been the blurring of the distinction between active and passive smoking. This is a tacit recognition of the fact that the data concerning passive smoking are weak, and the magnitude of the risk associated with ETS exposure uncertain.[24] But if exposure to ETS is drastically reduced, this will have the beneficial side effect of reducing opportunities for smokers to smoke and will thereby have a palpable effect on the rates of smoking-related disease. This objective was implicit in the strategy of the antismoking movement since the 1970s. As Ronald Bayer and James Colgrove, researchers at the Columbia University's Mailman School of Public Health, have recently argued, the focus on ETS was in part necessitated by the traditional hostility in American political culture toward public health interventions that are overtly paternalistic.[25] In other words, smokers didn't want to be told by the government not to smoke. This meant that if it could be shown that smoking had a harmful effect on the health of bystanders, this would provide a more effective, if indirect, means of curtailing smoking.

The linkage between restricting ETS and curtailing active smoking is often referred to in epidemiologic studies but only in a matter-of-fact way; rarely is it acknowledged as Bayer and Colgrave do, that this linkage is politically expedient. For example, a recent paper presenting the combined results of two large studies of ETS concludes as follows: "The implications of reducing exposure to secondhand smoke, however, go beyond the prevention of lung cancer in nonsmokers, since such measures to reduce exposure to secondhand smoke also result in a decreased opportunity for smoking among active smokers and a subsequent reduction in active smoking levels."[26] Equally rare in the mainstream literature is any recognition of the fact that the blurring of the distinction between ETS and active smoking is convenient in view of the weakness of the scientific case regarding ETS. Passive smoking was too useful and too powerful a weapon in the tobacco wars for the antismoking camp to forego its use, even if this meant overstating the scientific evidence.

***

The EPA report did not go without criticism from a variety of quarters. Needless to say, the tobacco industry did its best to argue that the science was weak and that the case against ETS was really a vehicle for a crusade against tobacco. But the industry's record of past deception insured that any merit its arguments concerning ETS might have could be easily discounted. More significant was the fact that a number of scientists in the field of cancer research were either explicitly or implicitly critical of the report. These included Ernst Wynder, Nathan Mantel, Philippe Shubik, and Roger Jenkins. Wynder had conducted one of the first studies showing a strong association between cigarette smoking and lung cancer and devoted a long career to studying the effects of smoking and diet on disease; Mantel, a statistician, had made major contributions to epidemiologic methods; Shubik had done pioneering research on carcinogenesis; and Jenkins, an analytical chemist has devoted several decades to obtaining accurate measurements of actual exposure to ETS. It was also noteworthy that the presumably impartial Congressional Research Service concluded that the "statistical evidence does not appear to support a conclusion that there are substantial health effects from passive smoking." Surely, one of the more curious responses to the EPA report was a legal ruling against it and in favor of a number of tobacco companies in 1998 by a federal judge for the Winston-Salem district of North Carolina. Judge William Osteen concluded that the EPA had made selective use of the existing evidence in order to reach a preordained conclusion that ETS was a class A carcinogen.[27]

But skepticism about the EPA's methods and technical criticisms voiced by a small number of individuals had little discernable effect on the momentum of the antismoking forces' focus on ETS. For one thing, the isolated voices concerned with not overstating the science had nowhere near the clout of federal agencies and voluntary organizations with staffs that were highly motivated to disseminate the new dogma and influence policy. Second, one could only demonstrate the distortion of the facts by referring to specific studies and technical issues which were difficult to convey to nonspecialists. Third, in a climate in which the discussion of questions relating to tobacco and health were cast in Manichean terms, it was only too easy for the tobacco control activists to dismiss any criticism of their campaign by implying that critics must be in league with the tobacco industry. This provided the activists with a convenient tactic for avoiding any discussion of the underlying data. The tobacco industry's long history of mendacity and distortion provided the activists with an all-purpose alibi attesting to their own righteousness. The report spoke to the underlying concerns of a growing proportion of the population, who, as surveys made clear, felt that exposure to ETS endangered their health. And, as the EPA had correctly anticipated, its report provided a major impetus for new and tougher ordinances restricting smoking in the workplace and in public places. In the years since publication of the EPA report a massive regulatory environment has evolved to extend restrictions on smoking and to administer policies regarding indoor air quality.[28] As a result of these policies enacted at the federal, state, and local levels, exposure to ETS has been dramatically reduced.[29]

Empowered by the EPA report, the antismoking forces set their sights on what they contended was a much larger source of ETS-induced death, namely heart disease. But before turning to this topic, we must look at the findings from environmental scientists and analytical chemists who tried to obtain accurate measurements of how much smoke people were actually exposed to.

MONITORING ACTUAL EXPOSURE TO ETS

Up until now, we have focused on epidemiological studies of long-term health effects of exposure to passive smoking. However, studies that have attempted to make rigorous measurements of the level of pollutants that people breathe in in smoky environments provide another essential piece of the picture. In the mid-1970s, two researchers at the Harvard School of Public Health, William Hinds and Melvin First, carried out careful measurements of environmental tobacco smoke in public places, such as restaurants, bus terminals, and bars in order to evaluate the health implications for nonsmokers.[30] They took great care in making their measurements, focusing on airborne nicotine, whose sole source is tobacco. Their paper, published in the *New England Journal of Medicine*, reported that at the measured concentrations, exposure to environmental tobacco smoke in public places was equivalent to smoking about 0.004 (4 thousandths) of a cigarette per hour. For a person who was exposed 5 hours a day, 7 days a week, and 52 weeks a year at this level, this would be the equivalent of smoking about seven cigarettes per year. It should be remembered that in the 1970s, when this study was carried out, there were few restrictions on smoking.

Hinds and First's work dates from the early 1970s, when environmental tobacco smoke was beginning to attract attention as a health concern. Dr. First told me that his involvement with this issue "began as a personal crusade against smoking." "I wore at all times a large lapel button that said,

'I'm for clean air. Don't smoke!' I had these buttons made by the hundreds at my own expense, and whenever someone of like mind asked where they could get the pin, I would reach into my pocket and present them with one."[31] Their work had been undertaken with financial support from the Massachusetts Lung Association at a time when First was a member of its governing board. The researchers' expectation had been that they would find "a severe health risk." However, when their results failed to confirm the original hypothesis, the lung association was not inclined to release their report. Fortunately, as First told me, Harvard University had a firm policy that made research results the property of the researchers, so they were free to publish.

In spite of the fact that First had been motivated by concern for nonsmokers' right to breathe unpolluted air and that his goal was to do credible scientific work on this topic, his results were uniformly ignored. None of the surgeon general's reports ever mentioned the study. Although his *New England Journal of Medicine* paper is cited in the bibliography of the EPA report, its results are not discussed or given any weight. As he explained: "People had already made up their minds that smoking and smoke were bad and would not accept anything to the contrary as valid or useful. I was better accepted when I was peddling my buttons."

Since the publication of the EPA report, much more sophisticated studies measuring typical exposures to ETS have been carried out. The largest of these was conducted by Roger Jenkins and his group at Oak Ridge National Laboratories in Tennessee.[32] In the early 1990s, Jenkins, an analytical chemist who has worked on environmental tobacco smoke since the 1980s and has written a comprehensive textbook on ETS,[33] decided that the major gap in the available data was assessment of ETS in residential situations. He wanted to do an adequate-sized study on ETS concentrations and personal exposure. When he informally approached the Department of Energy and the Environmental Protection Agency, neither institution showed much enthusiasm for funding such a study. So when the Center for Indoor Air Research (CIAR), an industry-funded organization that supports research on tobacco, asked him how he would like to do a study of personal exposure to ETS in both the home and the workplace, this fit right in with what he thought was the next step. The grant from CIAR enabled Jenkins and his coworkers to carry out the largest and most careful study of exposure to environmental tobacco smoke in everyday life.[34] Most previous studies had taken measurements using fixed air sampling devices and had relatively small numbers of subjects. But the use of stationary sampling devices does not tell you what a person's exposure is as he changes his location. Furthermore, most studies relied exclusively on cotinine (the major metabolite of nicotine) as a marker of the dose absorbed by the individual. However, Jenkins showed that cotinine levels correlate poorly with airborne nicotine measurements, indicating that it is an imperfect marker of exposure.

Jenkins' group enrolled a sample of one hundred nonsmoking adults in each of sixteen U.S. cities, yielding a total study population of 1,600. Each participant collected a sample of air from his, or her, breathing zone over a twenty-four hour period, using two separate pumps, one for the hours at work and the other for the remaining hours of the day (fig. 6.1). Participants also recorded their movements over the course of the day and noted whether, for example, restrictions on smoking in their workplace were strictly adhered to or not. The researchers also built a variety of safeguards and quality control checks into the study to ensure that the data they collected were of the highest quality. The air samples were analyzed for eight different compounds, including particulate matter and nicotine.

The results of Jenkins's study can be summarized as follows. First, subjects who live in smoking environments are exposed to 30–60 times more ETS than subjects living and working in nonsmoking environments. Second, ETS levels are lower than predicted from previous studies that used shorter duration of measurement or area measurements. Third, ETS levels determined in smoking homes are comparable to slightly lower than those determined for previously reported residential studies. Finally, the home appears to be a more important source than the workplace for ETS exposure, owing to the time spent in the home environment. In fact, exposures of subjects whose spouse smoked unrestrictedly within the home were from 2 to 4 times higher than those of subjects who worked in locations where smoking occurred and was not restricted. Typical workplace exposures were only a fraction of those estimated by the Occupational Safety and Health Administration. Commenting on his results in an e-mail exchange, Jenkins said that "the primary source of exposure these days seems to be the home. . . . If your workplace is one in which smoking is unrestricted, but you live with a smoker, you are likely still to get more exposure outside of work."[35]

When I asked Jenkins to give a rough estimate of how many cigarettes the average person's exposure to ETS would be equivalent to, he said he thought this could be done in a scientific way but that one had to be careful

FIGURE 6.1 Air sampler used in the Sixteen Cities Study.

about one's assumptions. He estimated that over the course of a year, mean levels measured in his study would suggest that women who had never smoked but lived with a smoker might receive about eight to ten cigarettes worth of nicotine and particles from ETS exposure. This estimate is in good agreement with results from a study similar to Jenkins' which was carried out in the 1990s in the U.K.[36]

Based on his knowledge of the studies which have measured ambient exposure to ETS among nonsmokers in the home and in the workplace, Jenkins has two fundamental criticisms of the EPA report. First, the EPA assumed a very low rate of smoker misclassification of 1.09 percent. (This refers to the fact that in the epidemiologic studies, some of those who were classified as never smokers were actually smokers. If ever smokers are included in a study of secondhand smoke exposure in never smokers, this will produce a spuriously elevated risk estimate, due to the fact that active smoking is a strong risk factor for lung cancer. Since some degree of misclassification of active smokers as never smokers is known to occur in epidemiologic studies, it is necessary to correct for this misclassification.) From his own work and that of others, Jenkins argues that a rate of 3 percent misclassification rate would be more in line with reality. This would have reduced the EPA's estimate of the relative risk for never smoking women married to a smoking spouse from 1.19 to close to 1.10, and it would no longer have been statistically significant.

The second adjustment used by the EPA that Jenkins questions was the adjustment for "background exposure to ETS." Since exposure to a spouse who smokes is just one component of total ETS exposure, the EPA reasoned that their estimate should be adjusted upward by some amount based on other sources of exposure, including that in the workplace. However, according to Jenkins, background levels of ETS are much lower than what the EPA suggested, and he had much more solid data from his Sixteen Cities Study to address this question. The EPA defined a "Z factor" as the ratio of exposure to ETS from spousal smoking plus other sources to ETS exposure from nonspousal sources alone and used a Z factor of 1.75. Jenkins considers that actual number should be between 8 and 25, reflecting the much larger contribution of ETS exposure from a smoking spouse relative to other sources of exposure. Both of these corrections to the EPA's numbers would act to substantially reduce the estimate of the risk of lung cancer due to spousal smoking.

"When you do the math, it is really amazing how low most typically encountered ETS concentrations really are. ETS particles at 200 $\mu g/m^3$ are about 0.175 parts per million. Most commonly encountered nicotine concentrations are on the order of one-half to a few parts per billion. NNK at 4 ng/m^3 is about 0.5 parts per trillion, BAP about the same. When you get down to the sub-parts per billion, indoor air is a very complicated soup. So many of the same substances found in ETS are present in indoor air from other sources at comparable or higher concentrations."[37]

On the basis of his own calculations, Jenkins believes that the EPA appears to have overestimated the number of lung cancer deaths by about a factor of 10. Rather than three thousand per year, three hundred may be closer to the mark. Of course, the problem is that this small an effect is difficult to detect using epidemiology.

ETS AND HEART DISEASE

Once the case was made that ETS was linked to lung cancer in nonsmokers, it was logical for epidemiologists to turn their attention to a possible association of secondhand smoke and coronary heart disease, the leading cause of death in the United States. Although it is not widely appreciated, active smoking actually kills more people through coronary heart disease than through lung cancer. This is so in spite of the facts that (1) the association of active smoking with heart disease is much weaker than with lung cancer (the relative risk for current active smoking with heart disease is about 2.0 compared to a relative risk for lung cancer of about 20.0), and (2) smoking is just one of a number of cardiovascular disease risk factors, whereas it is by far the predominant risk factor for lung cancer. The reason for the greater number of deaths from heart disease attributable to smoking is that heart disease is a much more common disease than lung cancer. Even a risk factor with a modest relative risk will cause a large number of cases if the disease is common and the prevalence of the factor (i.e., smoking) in the population is substantial. And since active smoking is an important cause of heart disease, it made sense to examine the effects of ETS on the leading cause of illness and death in the United States. However, it should also be noted that the death rate from heart disease declined dramatically throughout the second half of the twentieth century, both in smokers and nonsmokers.[38, 39]

By the late 1990s, at least eighteen epidemiological studies had appeared examining the relationship between exposure to environmental tobacco smoke and risk of heart disease, and several meta-analyses were published summarizing the results.[40] These meta-analyses indicated that exposure to secondhand smoke was responsible for a roughly 25 percent increase in the risk of coronary heart disease in nonsmokers. And application of this level of increased risk to the U.S. population yielded estimates of 35,000–40,000 deaths in nonsmokers per year attributable to ETS.[41] These findings were widely accepted. However, as was the case with the lung cancer data, here too, rather than acknowledging the weaknesses and inconsistencies in the data, certain findings were selected and overstated, and studies that did not find an effect were disregarded.

Methodologically, the most important studies on the question of ETS exposure and heart disease were cohort studies, which examined the occurrence of death from myocardial infarction (heart attacks) in nonsmokers married to smokers as compared to nonsmokers married to nonsmokers. As in the case of lung cancer, meta-analysis was used to take a weighted average of the results of individual studies in order to obtain a "summary estimate" that theoretically would be the best estimate of the association of ETS exposure with heart disease. However, meta-analysis is a technique that can easily be misapplied. In an editorial accompanying one of the meta-analyses, the statistician John Bailar III articulated a number of questionable assumptions applying to these meta-analyses.[42] I will focus on just two points concerning the credibility of these highly influential and widely accepted meta-analyses. First, a cardinal rule of meta-analysis is that all relevant studies be included. To exclude studies on subjective grounds is to run the risk of distorting the outcome of the procedure. Second, the selection of results from each study for inclusion in the meta-analysis should be consistent from one study to the next.

Regarding the first point, because of their enormous size, the two ACS prospective studies would have a preponderant influence. For this reason, the question of whether results from both studies should be included in a meta-analysis became a pivotal consideration. Two analyses using data from the American Cancer Society's cohort studies Cancer Prevention Study I (CPS I) and Cancer Prevention Study II (CPS II) had been published in the 1990s. One was conducted by two tobacco industry scientists LeVois and Layard using data from both CPS I and CPS II;[43] the other analysis limited to CPS II was carried out by the American Cancer Society itself.[44] Both studies were based on hundreds of thousands of never smokers. The first paper by the industry consultants reported no association with exposure to a spouse who smoked, while the second paper, by the American Cancer Society, presenting an analysis of CPS II, reported very weak results, with little evidence of increased risk in the groups with the highest exposure. All three meta-analyses included the paper from the ACS analyzing CPS II but excluded the CPS I as analyzed by LeVois and Layard for the apparent reason that, if it were included, this large null study would have substantially reduced the overall relative risk in the meta-analysis, owing to its large numbers.

A major justification given by the ACS for excluding the LeVois and Layard analysis is that because there were no restrictions on smoking in the 1960s, when CPS I was initiated, virtually everyone was exposed to secondhand smoke. And this, they contended, renders the study of no value for addressing the issue of secondhand smoke. However, these assertions by ACS do not stand up to scrutiny. First, exposure to secondhand smoke, although widespread, was not "ubiquitous" in the 1960s and 1970s, as the ACS claims. Based on ACS's own data, exposure to a smoking spouse is the most important source of exposure, and in the sixties, roughly 15 percent of males were nonsmokers and 26 percent were former smokers.[45] In addition, the majority of women enrolled in CPS I were housewives, so that exposure to secondhand smoke at work is not an issue in that group, which could be analyzed separately. Furthermore, the ACS saw fit to include other cohort studies conducted during the same time period in their meta-analysis. So it would appear that ACS had no objection to including studies from the 1960s and 1970s, so long as they showed the desired result. Finally, it will be recalled that Lawrence Garfinkel, a former head of epidemiology at ACS, had seen fit to use CPS I data to address the effects of ETS exposure.

Concerning the second point, the authors of the three meta-analyses did not always apply the same criteria in selecting the results to be included in the meta-analysis. In fact, in the American Cancer Society's analysis, where there was a choice between two numbers, the meta-analysts generally chose the higher of the two[46] (table 6.1).

One frequently cited meta-analysis is that of Law and Wald that appeared in *BMJ* in 1997.[47] After reading this paper you feel that there could not be a shred of doubt that ETS exposure is associated with a 25 percent increased risk of coronary heart disease in nonsmokers. This type of headlong certainty should give one pause at the outset. In spite of the authoritative tone of their piece, which brooks no contradiction, their conclusion rests on several questionable arguments. They repeat the arguments made in the ACS analysis and also have no problem justifying exclusion of the analysis of CPS I data. They also put great emphasis on short-term clinical and laboratory studies, indicating that even brief exposure to ETS can affect cardiovascular indices, such as platelet aggregation and the lining of blood vessels. This may be true, but these are only short-term effects, and their relationship to mortality from heart disease is not known. In any other context, the tenuousness of the supporting arguments would be acknowledged. In any event, the fact that there are speculative biological mechanisms supporting a possible effect of ETS exposure on cardiac health is not a justification for overstating the epidemiological evidence.

This is where the consensus concerning ETS and CHD stood in 2003 when James Enstrom and I published a detailed paper on the question using the California portion of CPS I, the same study that Garfinkel had analyzed back in 1981.

THE *BMJ* AFFAIR

The paper, which appeared in May 2003 in the *British Medical Journal*, or *BMJ*, reported the results of a large study of environmental tobacco smoke exposure and mortality from lung cancer, heart disease, and chronic lung disease.[48] This publication was the result of years of work by Enstrom, a research professor at the UCLA School of Public Health who had been conducting epidemiologic studies of cancer for nearly thirty years. In 1991, Enstrom was given access to data on the 118,000 California subjects in the American Cancer Society's Cancer Prevention Study I (CPS I) of one million Americans. This occurred because much of Enstrom's own epidemiologic research on California Mormons had been funded by the ACS since 1973 and he had formed a close working relationship with two successive directors of epidemiology at the society, Lawrence Garfinkel and Clark

TABLE 6.1 Overall Relationship Between ETS Exposure and CHD Mortality in U.S. Cohort Studies

Group and Study	Relative Risk by ETS Exposure Category[a]		
	RR (Former/Never)	RR (Current/Never)	RR (Ever/Never)
Males			
A. Svendsen et al. (1987)		**2.11[e] (0.69–6.46)[e]**	
A. Butler (1988)—AHSMOG			**0.55[d,e] (0.31–0.99)[e]**
A. Sandler et al. (1989)			**1.31[b] (1.05–1.64)**
A. Steenland et al. (1996)	0.96[b] (0.83–1.11)	**1.22[b] (1.07–1.40)**	1.09[b,d] (0.99–1.21)
B. Enstrom and Kabat (2003a)	0.94 (0.78–1.12)	0.94 (0.83–1.07)	0.94 (0.85–1.05)
C. LeVois and Layard (1995)	0.95 (0.83–1.09)	0.98[d] (0.90–1.06)	0.97 (0.90–1.05)
Females			
A. Garland et al. (1985)	3.00 (0.8–12.0)[c]	2.25 (0.5–11.0)[c]	**2.73[c] (0.7–11.0)[c]**
A. Butler (1988)—spouse pairs	0.96 (0.55–1.66)	**1.40 (0.51–3.84)**	1.05[b] (0.64–1.70)
A. Butler (1988)—AHSMOG			**1.51[d] (0.99–2.29)**
A. Sandler et al. (1989)			**1.19[b] (1.04–1.36)**
A. Humble et al. (1990)		**1.29 (0.79–2.10)**	
A. Kawachi et al. (1997)		**1.87 (0.56–6.20)**	
A. Steenland et al. (1996)	1.00[b] (0.88–1.13)	**1.10[b] (0.96–1.27)**	1.04[b,d] (0.95–1.15)
B. Enstrom and Kabat (2003a)	1.02 (0.93–1.11)	1.01 (0.93–1.09)	1.01 (0.94–1.08)
C. LeVois and Layard (1995)	0.99 (0.93–1.05)	1.05[d] (1.01–1.09)	1.03 (0.98–1.08)

Source: Enstrom and Kabat, 2006.

[a]In most studies exposure is based on the spouse's smoking status.

[b]Multivariate-adjusted RR was used (otherwise, age-adjusted RRs were used).

[c]RR was approximated from available published data.

[d]RR was based on combining other published RRs.

[e]Bold RR was used in meta-analysis by Thun et al. (1999).

Heath. Since the ACS had only followed the CPS I cohort from its inception in late 1959 through 1972, Enstrom's objective was to extend follow-up of the cohort to 40 years in order to relate long-term mortality trends to active smoking, passive smoking, smoking cessation, and other risk factors. He had an impressive track record using large cohort studies to address important questions about the effects of such lifestyle factors as smoking and diet on the risk of chronic disease. Given his long-standing interest in the causes of lung cancer in nonsmokers,[49] it was natural for him to want to study the effects of exposure to secondhand tobacco smoke. There was great interest in this subject following publication of the EPA report in late 1992, when the campaign to use smoking bans to reduce exposure to secondhand smoke and to restrict smoking was in full swing.

In 1991, Enstrom obtained initial funding from the University of California Tobacco-Related Disease Research Program (TRDRP) to examine active smoking, passive smoking, and smoking cessation in relation to long-term mortality in CPS I. The TRDRP had been established by the California legislature in 1989 to support "research efforts related to the prevention, causes, and treatment of tobacco-related diseases" and was funded by a 25-cent per pack cigarette surtax. Despite impressive success in determining the current vital status of subjects last contacted in 1972 and in getting survivors to respond to a 1994 follow-up questionnaire, Enstrom was denied continued TRDRP funding necessary to complete his study. Peer reviewers felt that this cohort had "outlived its usefulness" and would not add much to the understanding of the health effects of either active or passive smoking. Enstrom was also unable to obtain any funding from the ACS. However, in 1998 he was able to obtain the funding to complete follow-up and examine passive smoking and mortality from the Center for Indoor Air Research (CIAR), a research organization concerned with indoor air issues and funded primarily by the tobacco companies.

When Enstrom asked me in 1997 whether I would be interested in serving as a consultant on epidemiology relating to secondhand smoke on his proposed CIAR analysis, I agreed, because I felt that using long-term follow-up of the California portion of CPS I to study the impact of spousal smoking on mortality was a worthwhile undertaking. I knew Enstrom from his published work on Mormons and lung cancer in nonsmokers and considered him a careful scientist.

The study comprised over 118,000 adults enrolled by the American Cancer Society in 1959 who were followed until 1998. Of these participants, nearly 36,000 were never smokers who had a spouse in the study with known smoking habits. In 1999, a two-page questionnaire was sent to surviving members of the cohort to obtain information about their exposure to other people's tobacco smoke at home, work, and in other settings, as well as to verify their own status as never smokers and other personal information. This information was obtained from 3,100 of the 8,700 members of original 36,000 nonsmokers who were alive in 1999—an impressive feat after 39 years.

The results of this large study with almost 40 years of follow-up showed no evidence of an elevated risk of coronary heart disease or of lung cancer in either male or female nonsmokers with spousal ETS exposure. There was, however, suggestive evidence of a slight association with nonmalignant respiratory disease. Following a detailed presentation of the data on spousal smoking in the paper, we showed the impressive and quite regular dose-response relationship of active smoking with the three diseases under study, in both males and females. Even current smokers of 1–9 cigarettes per day had statistically significantly elevated risks for all three conditions in both males and females. We presented these results in order to contrast the powerful and more certain effects of active smoking with those of ETS exposure, something that is all too rarely done. We also demonstrated that the study had the ability to pick up a real effect, even for light smokers and former smokers.

After two rejections of the paper by American journals, we submitted it to the *BMJ*, where, to our surprise, it got a rigorous yet very favorable review. We believe that this occurred because the *BMJ* editors examined our paper very closely and fairly and had an editorial policy in place not to automatically reject research funded by the tobacco industry but rather to review the work on its merits. Three reviewers and the editors felt that the paper was important and should be published. As is normal, the reviewers made suggestions for how the paper could be improved. One of the reviewers in particular made a number of insightful and creative suggestions for how to provide the fullest and most rigorous analysis of the available data. We carefully responded to the reviewers' comments, and, as a result, the final version of the paper was greatly strengthened. This is an example of the enormous strength of the peer review process.

The paper appeared in the May 17 issue of the *BMJ*. Unbeknownst to us, the editors had made it the cover story under the catchy headline: "Passive Smoking May Not Kill." An accompanying photograph on the cover showed the entrance to a California office building prominently displaying a notice warning of the adverse health effects of passive smoking. This high-profile treatment of the paper, as it turned out, generated a huge and immediate uproar on both sides of the Atlantic, which neither Enstrom or I had been prepared for.

Two days before the *BMJ* issue was officially available, the American Cancer Society called a press conference at which its Vice President for Epidemiology and Surveillance Research, Michael Thun, attacked our study as industry-funded and fatally flawed.[50] At the same time, the head of the British Medical Association, the parent organization of the *BMJ*, blasted our paper, the journal, and its editor-in-chief for publishing it.[51] What was most noteworthy was that these attacks took place *before* the full version of our paper was even available, so that none of these outraged critics could have read through the 3,000 words of text and the ten tables and come to a judgment on the merits of the paper. But clearly that was not the point. The point was to nip in the bud and to irreversibly discredit a major paper reporting the results of a very large study associated with the name of the American Cancer Society that came to the unacceptable conclusion that, at least in this particular population, there was no suggestion of an association between spousal smoking and either coronary heart disease or lung cancer.

In the weeks that followed the publication of the paper, the journal's Web site was flooded with "rapid responses" to our paper. Anyone who desired to could write an electronic rapid response that would be permanently posted on bmj.com. (A small number of these letters were selected several months later to be published in the print edition[52]). Over a period of several months, 140 letters appeared on the Web site. The majority of the writers appeared to have taken their cue from the American Cancer Society's and the British Medical Association's attacks and vented their outrage and indignation that the prestigious *BMJ* would publish such a scurrilous piece of industry-sponsored trash. Many of these writers were associated with antismoking organizations or were health professionals. A sizable minority of letter writers applauded our paper and some attacked the antitobacco activists for being hypocritical about industry funding, since, it was claimed, they were in many cases the recipients of substantial funds from "Big Pharma." Thus, for the vast majority of responses, our paper was merely an occasion for voicing their deeply held views in the "tobacco wars." While our null results elicited very strong reactions, very few of the responses showed any familiarity with, or interest in, the details of the study or how it compared with previous studies. In fact, only 3 percent of the rapid responses made any reference to the specific data in the long and complex paper. When selected correspondence was published at the end of August 2003, the editors defended their decision to publish the paper, pointing out that the attacks had failed to acknowledge the considerable strengths of the study in addition to its weaknesses, since all studies on this topic have considerable shortcomings.[53] They further pointed out that the correspondence had been characterized more by its intense emotion than by dispassionate analysis. It should also be noted that not one epidemiologist pointed out that the issue of passive smoking had been overblown. Nor did a single commentator point out that our findings were not very different from those of other studies on this question.

Amidst the fury of denunciation of our paper by the ACS and the antismoking forces, a number of crucial distinctions got glossed over. The fact that the last seven years of a 39-year project were funded by the tobacco industry was used to imply that both the researchers and the study were tainted. This was apparently enough to discredit the results in the minds of many people who did not stop to examine the matter more closely. Although no one made the charge, the implication was that we had somehow twisted the data in order to obtain a null result. In point of fact, as we stated in the published note on potential conflicts of interest accompanying the paper, the tobacco industry did not participate in any way in the analysis and did not see the paper prior to publication. Second, the contention that CPS I could not be legitimately used to address the question of passive smoking because "in the 1960s everyone was exposed" is demonstrably false—particularly in the case of housewives who constituted 60 percent of the women in CPS I and for whom spousal smoking could be expected to be a major contributor to total ETS exposure.[54] Third, as mentioned, earlier Thun himself had included in his meta-analysis two U.S. cohort studies from the 1960s, suggesting that there was no objection to including studies that measured exposure in the 1960s so long as they showed the desired result. Fourth, as noted earlier, Lawrence Garfinkel, Thun's predecessor at ACS, had felt it was legitimate to use CPS I to study passive smoking. Finally, and most significantly, ACS had the data for the time period 1959–1972 and could easily have redone the analysis and identified any error in our paper. If they have in fact carried out the analysis, as some suspect, they have decided not to publish the results that were not to their liking.

Ours was only one study, and we made no special claims for it, only that it be judged by the same standards as the other studies on this topic. If this is done, one has to acknowledge that our paper has a larger number of heart attack deaths than any of the other studies included in the meta-analyses mentioned above; it also has spousal smoking information obtained at several points in time; and it is one of very few papers to present information on misclassification of exposure. Not one of our critics acknowledged these strengths. However, perhaps most revealing is the similarity of our findings, particularly for heart disease with the findings from other U.S. cohort studies and, in particular, those from the American Cancer Society cohort studies. Putting our results side by side with those of LeVois and Layard and those of Steenland, the estimates are quite similar.[55] None of these large analyses showed any hint of a dose-response relationship with increasing exposure or any clear excess in the highest exposure group. For women—the group that should be most informative about the effects of secondhand smoke—the results were particularly weak. Table 6.2 compares the results for females from Enstrom and Kabat with those from the ACS reported by Steenland et al.

When the results of individual studies are selected according to consistent criteria and when results from CPS I are included in the meta-analysis, the overall association of exposure to spousal smoking with heart disease is shown to be on the order of a 5 percent increase in risk (that is, a relative risk of 1.05) rather than a 25 percent increase (relative risk of 1.25).[56] This

TABLE 6.2 Association of Spousal Smoking with Lung Cancer in Female Never Smokers in CPS I and CPS II

	Enstrom (CPS I)		Steenland (CPS II)	
	RR	95% CI	RR	95% CI
Spousal smoking				
Never	1.00	(reference)	1.00	(reference)
Former	1.02	(0.93–1.11)	1.00	(0.90–1.48)
Current				
1–19 cpd	1.07	(0.96–1.19)	1.15	(0.90–1.48)
20 cpd	1.04	(0.92–1.16)	1.07	(0.83–1.40)
21–39 cpd	0.95	(0.80–1.06)	0.99	(0.67–1.47)
40+ cpd	0.83	(0.65–1.06)	1.04	(0.67–1.61)
Current total	1.01	(0.93–1.09)	1.10	(0.96–1.27)
Ever	1.01	(0.94–1.08)	1.04	(0.95–1.15)

Source: Enstrom and Kabat, 2006.

smaller increase is consistent with the estimates derived from studies using personal monitoring. If this lower estimate of the association between secondhand smoke exposure and heart disease is accurate, this would mean that the figure of 40,000 or 50,000 deaths from heart disease attributable to ETS is greatly exaggerated.[57]

RECENT AGENCY REPORTS

In the past few years a new round of voluminous reports on the health effects of secondhand smoke exposure has been issued by International Agency for Research on Cancer, the California Environmental Protection Agency, and the U.S. Surgeon General.[58] These reports represent a continuation of the approach taken in the 1992 EPA report—that is, to include new studies that have appeared and to update the summary risk estimate characterizing the association of ETS exposure with specific diseases. While they have the appearance of judicious and comprehensive summations of the available scientific evidence, they are clearly designed to support the ever-expanding policies of restricting smoking. And, in view of this goal, they give a particular slant to the evidence, overstating the quality of certain types of data and neglecting other relevant evidence. They are meant to persuade by "piling up the evidence," as John Bailar put it.[59] But, again, they studiously avoid putting ETS exposure in perspective and acknowledging weaknesses and inconsistencies in the individual studies. It is clear that on the subject of tobacco the accepted practice of a comprehensive and critical review of all of the evidence does not apply. Having said this, there are some noteworthy differences among the three reports.

The IARC report is the most restrained of the three reports, concluding that the evidence is "sufficient to conclude that involuntary smoking is a cause of lung cancer in never smokers."[60] But it refrains from estimating the number of deaths attributable to ETS exposure, leaving open the question of how important passive smoking is as a cause of fatal disease.

The most extreme of the three reports is that of the California EPA, which represents the first step in a process that could lead to the regulation of ETS in the outdoor environment in the state of California. Regulation of smoking in outdoor areas where people congregate would, of course, further reduce the public space in which smokers can indulge their habit and would represent a bold new step in the progression that started with the first modest restrictions on smoking in the 1970s. Seen in this way, it is entirely logical. But how is the available science used to support this initiative?

Rather than presenting all of the evidence and impartially assessing its strengths and weaknesses, the report emphasizes the positive findings. At the same time, little attention is devoted to the inconsistencies and limitations of the evidence. Motivated by the drive to justify this ambitious undertaking by uncovering yet new diseases that can be linked to ETS, the authors of the report recklessly concluded that premenopausal breast cancer is linked to passive smoke exposure. Since even active smoking is not convincingly linked to breast cancer, this step by the California EPA has caused consternation even among scientists who are activists on the effects of ETS. Along with its selective approach to the evidence, like the other two recent reports, the California EPA report studiously fails to make basic distinctions that would help put the issues under discussion in perspective.[61]

The 2006 U.S. Surgeon General's report appears more judicious and rigorous. However, it too fails to acknowledge relevant findings that would help put the passive smoking issue in perspective, such as the fact that lung cancer rates in never smokers have remained flat over the decades when the effects of active smoking were making themselves felt; the fact that a large autopsy study showed virtually none of the advanced pathologic changes typical of smokers' lungs in the lungs of never smokers; and the fact that average ETS exposure is orders of magnitude lower than that of the active smoker to tobacco smoke. The Surgeon General's report also overstates the quality and consistency of the epidemiologic evidence on passive smoking. Tellingly, it accepts the California EPA estimate of 50,000 deaths per year in the United States attributable to ETS exposure. As demonstrated above, this estimate (due mainly to deaths from heart disease) is based on slanted and selective meta-analyses. Inclusion of results from the large CPS I would have drastically reduced the estimate of the relative risk and, thereby, the number of deaths attributable to secondhand smoke exposure. Although the Surgeon General cited the more recent California EPA report of 2005, it saw fit to exclude from consideration the 2003 paper Enstrom and Kabat.

Although respected scientists have played a role in their redaction, these documents themselves bear the unmistakable stamp of the purpose for which they were written—to further bolster the consensus in order to affect policy. It is unfortunate that those with responsibility for planning and overseeing these reports judged that promoting a strong policy required overstating the scientific evidence. The weight of these documents is such that any divergent findings are judged to be inadmissible. Any criticism of their approach can easily, and cynically, be discounted as coming solely from the tobacco industry. Or, alternatively, it can be simply ignored. In this way a doctrine is built up that is antithetical to the openness that is a precondition for scientific discourse. What is objectionable is that these voluminous and authoritative-appearing documents convey the message that passive smoking is a major cause of fatal disease, which few scientists believe to be the case.

THE NEW MCCARTHYISM IN SCIENCE

Any discussion that seeks to understand the role that "social" and ideological factors can play in influencing the interpretation of the science concerning passive smoking must take note of the present-day climate prevailing in this field. Since the early 1990s, the whole area of antitobacco research, education, and policy has grown into a special domain unto itself that goes by the name of "tobacco control." A journal of the same name was established in 1992. Owing to the increasing availability of ample funding from foundations such as the Robert Woods Johnson Foundation and voluntary organizations like the American Cancer Society, the California state tax-funded Tobacco Related Disease Research Program, and especially the American Legacy Foundation, established as a result of the 1998 Master Settlement Agreement between the four largest tobacco companies in the United States and forty-six states, the field has grown enormously and has become a veritable industry.

It needs to be understood that in order to obtain funding in this area and justify how those funds are used, there is a powerful incentive to accept as a given the reigning consensus opinion on the health effects of passive smoking and a no less strong disincentive against drawing attention to the weaknesses or uncertainties in the underlying science. In the 1980s and earlier, before the rise of tobacco control as a field, it was almost exclusively epidemiologists who conducted studies of the health effects of tobacco exposure (I am talking here about nonlaboratory science). These epidemiologists typically conducted research on a wide range of diseases and exposures. They were true generalists. Tobacco was not their sole focus. In contrast, the new generation of tobacco control researchers included behavioral scientists, psychologists, health promotion specialists, and others, who, generally speaking, were more concerned with interventions to change behavior and with education and policy issues relating to smoking. They also tended to be more activist, since, as articulated by the Centers for Disease Control, a major goal for improving the nation's health was to reduce the prevalence of smoking-related disease.

A second key point to understand is that very few people are familiar with the original epidemiologic studies of ETS and lung cancer, and even fewer have subjected these studies to critical scrutiny. This being the case, most tobacco control researchers rely on the conclusions of consensus reports, which provide the justification they need to pursue their work. In addition to nonepidemiologists, there are also several key figures who are epidemiologists who have embraced tobacco control and who have played a major role in shaping the new consensus and have conferred on it the mantel of authority.

In this climate, in which so much is at stake—both in terms of funding and professional advancement, as well as in terms of ideology and public health goals—anyone who takes the science seriously and wants to assess its strengths and weaknesses is viewed as a threat to be neutralized. This situation has given rise to extraordinary attacks on the integrity of established scientists whose only documentable fault is to report findings in peer-reviewed journals and bring a critical (i.e., scientific) attitude to bear on the question of the long-term health effects of ETS exposure.[62] What is most extraordinary is that these attacks on established and respected researchers rely exclusively on the imputation of guilt by association and on alleged conflicts of interest. Because of the highly polarized climate and the self-proclaimed righteousness of the tobacco control position, those who engage in such attacks are not required to supply any evidence of wrongdoing or fraud. The fact that a researcher conducted research at a prominent university or national laboratory and that the research was partially supported by tobacco industry funding, even though the grant was administered by the recipient institution, is viewed by the activists as sufficient to discredit the targets of these attacks.[63] Having neutralized the offending research, the guardians of the consensus are then free to ignore it and pretend that it does not exist.

Back in 1993, the epidemiologist Ken Rothman wrote an article in the *Journal of the American Medical Association* entitled "Conflict of interest: The new McCarthyism in science."[64] Rothman, the author of one of the most respected textbooks of epidemiology, argued that human beings have complex motivations, and many different factors can influence both the work one undertakes and how one performs it. Financial conflicts of interest, he pointed out, are only one potential source of influence and should not be singled out as the only one. Other potential biasing factors include sexual orientation, religious beliefs, and a tenacious belief in specific scientific theories, since these too could conceivably influence their research and its interpretation. According to Rothman, it was mistaken and unfair to make judgments about a piece of work based on such "externalities" and to engage in speculation about motivation ("a task that ultimately requires psychic abilities"[65]). Objectivity, he argued, is not some privileged state of mind but rather is the result of vigorous and open discussion of different points of view. Rothman concluded that the work itself has to be judged on its merits. In other words, there is no substitute for the hard labor of critically evaluating a given piece of research—its methods, its data, its reasoning, and how these square with other comparable studies. Although his article preceded the new trend toward *ad hominem* attacks by some prominent tobacco control activists, I know of no better description of the mentality.

Only a small number of commentators have examined in any depth the wider influences that come into play in connection with the "tobacco wars." I have referred earlier to the work of Richard Kluger, Jacob Sullum, and Bayer and Colgrove. Two other contributions that offer valuable insights into the "sociology" of the tobacco wars deserve mention.

It took two sociologists—one a Canadian, the other a German—to bring out the deeper significance of the responses to the 2003 *BMJ* paper by Enstrom and Kabat. In a lengthy article, Sheldon Ungar and Dennis Bray analyzed the 144 rapid responses that appeared in the first two months following publication of the article, as well as newspaper coverage in Europe and North America.[66] The authors started from the premise that the scientific enterprise presupposes an openness to divergent ideas, even though contemporary science is increasingly subject to pressures stemming from "the commercial exploitation of science, expanded media coverage of research, growing public concern with risks, and the greater harnessing of science to social and policy goals." Ungar and Bray used the term "partisans" to describe individuals who not only have an unreasoned allegiance to a particular cause but who also engage in efforts to silence opposing points of view. Among their findings were that the rapid responses can best be understood as an attempt by partisans to silence results that go against the reigning consensus concerning the health effects of passive smoking. Drawing attention to the vehemence of many of the negative rapid responses, they commented that, "Silencing is based on intimidation, as partisans employ a strident tone full of sarcasm and moral indignation. There are elements of an authoritarian cult involved here: uphold the truth that secondhand smoke kills—or else!" They proceeded to examine the concern, voiced by many who attacked the paper, that it would attract extensive media coverage and set back the progress in achieving bans on smoking. But, contrary to this expectation, they found that the study received relatively little media attention and that only one tobacco company referred to the study on its Web site. Their explanation for this surprising lack of media coverage, given the strength of public opinion regarding smoking and passive smoking, was that the media engaged in "self-silencing." ". . . we suggest that doubts about the negative effects of passive smoke are inadmissible and unintelligible." And their overall conclusion was that "the public consensus about the negative effects of passive smoke is so strong that it has become part of a truth regime that cannot be intelligibly questioned."

Another figure who has taken an independent view of the reigning discourse concerning tobacco is Michael Siegel, a physician specializing in preventive medicine and a professor at Boston University. Siegel was a student of one of the prime movers behind the antismoking and clean air movements, Stanton Glantz of the University of California at San Francisco, and he has remained committed to the goal of protecting workers and others from exposure to tobacco smoke. However, a couple of years ago he decided that he had to speak out against what he saw as the increasingly objectionable tactics of the antismoking movement. These include misrepresenting the science on specific topics and an increasingly demagogic treatment of anyone who disagrees with their point of view. For the past two years, Siegel has written a blog entitled "The Rest of the Story: Tobacco News Analysis and Commentary,"[67] in which he tirelessly documents specific incidents of abuse or distortion and fills in the aspects of the question that have been ignored. His blog is a model of clarity and devotion to laying out all of the relevant evidence rather than citing selected or distorted facts to cudgel the "enemy." Some of the topics that Siegel has dealt with at length are the overstated claims of effects of smoking bans, the widely disseminated claim that as little as 30 minutes exposure to ETS can induce a heart attack in a nonsmoker, and the American Cancer Society's attack on Enstrom. It is interesting to note that both Ungar and Bray, and Michael Siegel find it useful to refer to the philosopher Michel Foucault's notion of a dominant discourse or "regime of truth" and to document the techniques that are used to suppress or silence dissenting points of view.

SIR RICHARD PETO BEFORE THE HOUSE OF LORDS

For decades during the twentieth century, the United Kingdom had among the highest rates of smoking and the highest rates of lung cancer in the world. As in the United States, a powerful antismoking movement grew up there, and in 2006 a wide-ranging law prohibiting smoking in "enclosed work spaces" was passed by Parliament. Public discussion of the effects of exposure to ETS has been extensive, and, as part of this debate, on February 14, 2006, the Oxford epidemiologist Sir Richard Peto testified before the House of Lords' Select Committee on Economic Affairs regarding the hazard posed by exposure to other people's tobacco smoke. Peto is one of the foremost epidemiologists in the world, who for the past three decades collaborated with Sir Richard Doll, until the latter's death in 2005. Among the problems Peto has been influential in addressing are the long-term effects of smoking in classic studies on up to fifty years of follow-up the British Doctors cohort study. In addition, in the early 1990s he was one of the first to draw attention to the aggressive marketing of cigarettes to non-Western countries, such as China, and to project the enormous future health toll from smoking-induced disease that could be expected in these countries. Peto has also used meta-analysis to combine data from many small clinical trials in order to address important questions, such as the effect of different treatment regimens for breast cancer. In all of his work, he is at pains to use available data to address questions that will affect large numbers of lives. Finally, he is the author, together with Doll, of the 1981 book *The Causes of Cancer*, which represents a magisterial effort to estimate the contribution of different causes to the incidence of the major cancers and to cancer as a whole. Given his background, Peto's testimony before the House of Lords on the question of the scientific rationale for a smoking ban in public places merits special attention. His choice of words and his characterization of the problem are striking and represent a point of view that has been rare in the discourse surrounding the health effects of passive smoking. Here is someone, who has devoted his career to using numbers and rigorous thinking to estimate the magnitude of important associations, refusing to be "quantitative" about the measurement of the risks due to environmental tobacco smoke.

> I am sorry, I know that is what you would like to be given, but the point is that these risks are small and difficult to measure directly. What is clear is that cigarette smoke

> itself is far and away the most important cause of human cancer in the world—that is, cigarette smoke taken in by the smoker—and passive smoking, exposure to other people's smoking, must cause some risk of death for the same diseases. Measuring that risk reliably and directly is difficult.
>
> The exposure that one would get when breathing other people's smoke obviously depends on the circumstances, but even heavy exposure would be something like one per cent of what a smoker gets, maybe in other circumstances 0.1 percent, so you would expect if there was just proportionality to get something of the order of 20 percent excess. That is what you see in the average of all studies, and people have pointed out the uncertainties—it could under-estimate the real hazards, it could over-estimate the real hazards. . . . You would expect an excess of a few percent, several per cent (for non-smokers).
>
> I do not want to be cast in the role of advocating banning smoking in public places or in private places. What I find concerning is that enormous risks, like the extent to which smokers kill themselves—there is about a 50 per cent chance that a person who smokes cigarettes and continues to do so will be killed by tobacco and that is just vastly greater than almost any other risks that there are around. *It is the relative causes of this that I somehow want to get across and we are concentrating now* [emphasis added—G.K.], because this is your task, on what are the effects of breathing other people's smoke, but the main way smokers kill people is by smoking themselves, not by killing other people—they are a lot better at killing themselves than they are at killing other people.[68]

It is clear that Peto does not think that the inevitably small risks associated with exposure to ETS can be quantified with any certainty, and that, therefore, it is more worthwhile to focus on larger risks that can be quantified.

***

Discourse concerning the health effects of passive smoking has two faces: one geared toward regulation and the public, the other purely scientific. The first approach entails making claims that are not qualified by any acknowledgment of methodological problems and giving estimates of deaths attributable to ETS based on slanted analyses. The second approach limits itself to the critical assessment of the existing evidence and attempts to put the findings regarding ETS in perspective. But this second face, confined to scientific journals and textbooks and not at all newsworthy or useful in the regulatory sphere, cannot compete with the first. The scientific consensus, as opposed to the ideological consensus, was recently articulated in a comprehensive review article on lung cancer occurring in never smokers published in journal *Nature*:

> Based on these data, the US National Institutes of Health (NIH) and the International Agency for Research on Cancer (IARC) have officially designated ETS as a human carcinogen. Nonetheless, the overall evidence indicates that ETS is a relatively weak carcinogen and most lung cancers in never smokers cannot be explained by ETS exposure alone.[69]

This succinct summation of the evidence contrasts with the increasingly voluminous compilations put out by agencies and designed to lay down an unquestionable doctrine. Those who appreciate its message are mainly scientists concerned with the causes of chronic diseases in people who have never smoked and the sector of the public that appreciates undistorted information. But the assertion of a major risk attached to passive smoking is much more compelling and satisfying than the acknowledgment that ETS is likely to be a minor cause of fatal disease in never smokers.

Box 6.2 Major Take-Home Points

There is no question that secondhand tobacco smoke contains the same carcinogens and toxins present in mainstream smoke or that exposure to ETS increases the incidence of certain acute respiratory conditions. Furthermore, there is no rational justification for anyone to be exposed to the unnecessary form of air pollution.

What is at issue here is how the science concerning ETS as a cause of serious chronic disease has been overblown and slanted.

During the 1970s the antismoking movement seized on the effects of exposure to environmental tobacco smoke on the bystander as an effective tool to combat smoking. This was *before* the appearance of any scientific studies on the long-term effects of such exposure.

When the first studies appeared in the early 1980s linking exposure to spousal smoking to lung cancer in never smokers, these studies were interpreted as providing evidence of an association. Further studies accumulated, and meta-analyses were carried out indicating that exposure to ETS conferred roughly a 20–25 percent excess risk of lung cancer in people who had never smoked. However, reports concerning ETS tended to minimize the considerable problems of reliably assessing a person's exposure and to avoid putting the effects of passive smoking in perspective by means of the obvious comparison with the effects of active smoking. In this way, the public was given the message that the effects of ETS were both more certain and more important than they actually were. The 1992 EPA represented a landmark in establishing the association of ETS with lung cancer beyond the possibility of doubt.

In the 1990s, studies appeared linking ETS to heart disease; however, here the association was more problematic because active smoking has a much more modest association with heart disease compared to lung cancer. Some influential meta-analyses used data selected in a biased manner and excluded studies that did not support the existence of an association, and skewed estimates of large numbers (35,000–50,000) cardiac deaths due to passive smoking have been widely accepted.

The question of the long-term health effects of passive smoking has, to a large extent, been removed from the realm of science. Even though, there are studies and reviews in reputable journals that convey a very different picture, these have little effect in the public realm, where the importance and certainty regarding health effects of ETS exposure cannot be questioned.

7

CONCLUSION

Epidemiology is so beautiful and provides such an important perspective on human life and death, but an incredible amount of rubbish is published.

—Sir Richard Peto, 2007

In the past, we had few research findings, while currently we have too many research findings. Therefore, getting rid of tentative-but-wrong research findings should become at least as important as finding new ones.

—John P. A. Ioannidis, 2005

Widespread confusion about what are important health risks is hardly surprising when one considers the positive din of information relating to health that assaults us on a daily basis—claims of benefits, supposed breakthroughs, policies issued by the federal government and professional bodies, and ever-multiplying threats. Although the media clearly plays a pivotal role in focusing attention on certain "stories," responsibility for this confusion cannot be ascribed solely to the media. Rather, it involves complex interactions between the producers and consumers of knowledge in the area of public health, with the participation of scientists, regulators, advocates, journalists, and others. In each of the case studies discussed in the preceding chapters, certain results were selected, highlighted, and publicized, whereas other relevant facts and considerations were ignored. These isolated findings then became counters in a form of ritual circulation among the different parties that gave them credence. Divorced from any context, these ostensibly scientific findings became things-in-themselves, and their circulation in the wider society appeared to give them solidity and to confirm them.

However, in the process of identifying and publicizing these supposed hazards, crucial distinctions were lost sight of. These include the distinction between *studying* a question and finding strong evidence of an effect, between association and causation, between the results of a single study and the totality of relevant evidence on a question, between strong evidence and mediocre or flimsy findings, between risk factors that have large and well-established effects and risk factors that are marginal or uncertain, or factors that simply merit more study. Finally, there is the distinction between science and politics.

Science, of course, is analytic and reductionist in its approach, but, as we have seen, a narrow focus can screen out relevant facts, which if acknowledged, might drastically affect the interpretation of the evidence. It would have been more difficult to make a case for DDT as a serious cause of breast cancer, if the "disconnect" between the decline in DDT levels in the environment and in human tissues, on the one hand, and the increasing incidence of breast cancer, on the other, had been acknowledged. It would have been hard to make a convincing case that the extremely low frequency electromagnetic fields from power lines and home appliances might constitute a major hazard if epidemiologists and health agencies had forthrightly acknowledged the lack of evidence that typically encountered fields have any documented biological effects at all. Similarly, it would have been difficult to make residential radon into an urgent and insidious threat if the magnitude of the risk of lung cancer posed by radon had been compared to that from smoking, and if the lack of convincing evidence of an effect in people who have never smoked had been acknowledged. Finally, the adverse effects of exposure to passive smoking would have been placed in a more rational framework if scientists and health agencies had made the obvious distinction between active smoking and passive smoking. Also, it would have been harder to maintain the widely publicized estimate of 50,000 heart disease deaths per year attributable to passive smoke exposure if the influential scientific papers had acknowledged such relevant facts as the steep decline in deaths from heart disease in both smokers and nonsmokers over the past fifty years, the low concentration of secondhand smoke compared to the smoke inhaled by the active smoker, and the general weakness and inconsistency of the results of epidemiologic studies.

The potential for isolated or limited findings to be transformed into major health hazards should alert us to the need for skepticism that extends both to the results from the latest study and to the "consensus" on certain high-profile topics. When confronted by claims of major risks or benefits, we need to keep in mind certain basic realities, such as the difficulty of elucidating the etiology of complex, multifactorial chronic diseases, like breast cancer, colon cancer, heart disease, and Alzheimer's disease; the hunger and impatience on the part of the public—that is, of all of us—for positive findings; and the role of extra-scientific factors and agendas in what gets studied and how results get reported. Too often, such fundamental facts of life are forgotten in the rush to judgment and the suspension of critical thought that accompany the manufacture of a new hazard.

That such indispensable distinctions and relevant facts get ignored in much of what passes for scientific discourse and reporting on science makes apparent the need for an explanation that steps outside of this process—the dissemination of selected, disembodied findings—and addresses the agendas and ideological stances that play a role in the hyping of certain hazards.

The focus on the latest, transient findings obscures the fact that most important advances in knowledge about factors that have profound effects on health are the result of a long and painstaking process of accumulation of evidence from observational, clinical, and laboratory studies. It took decades from the initial observations indicating that use of aspirin, and other nonsteroidal anti-inflammatory drugs, substantially reduced the risk of colorectal cancer to confirmation of this relationship in studies with long enough follow-up to detect an effect.[1] (While this knowledge does not presently lend itself to the prevention of colorectal cancer, and further research into the optimum dosage, length of administration, and side-effects are necessary, nevertheless, it represents a major step toward a preventive strategy.) Similarly, it took decades of work to establish the utility of SERMs (selective estrogen receptor modulators) in the treatment and prevention of breast cancer.[2] It is worth noting that these and other major advances have not received anywhere near the kind of attention that is devoted to marginal and uncertain effects.

Over time a process of winnowing takes place as certain findings get extended and confirmed and take on stability, and others—which may have appeared very promising—wither away or prove to be of marginal significance. A recent report by the World Cancer Research Fund provides a good example of this process.[3] The report represents the work of a team of scientists from nine universities who devoted five years to sifting through thousands of scientific articles on the effects of diet, physical activity, and body weight on different forms of cancer. In its main conclusions, the report confirms judgments of previous expert panels to the effect that obesity and alcohol consumption make clear contributions to the risk of certain cancers. A second conclusion based on more recent evidence is that a relatively high intake of red meat increases the risk of colorectal cancer. However, concerning the effects of intake of fat, fruits and vegetables, fiber, and vitamins and minerals, as well as concerning the effects of physical activity (apart from its protective effect on colorectal cancer), no firm conclusions could be drawn. This casts a rather different light on the myriad scientific reports on diet and cancer that have appeared over the years. In retrospect, one can see clearly that many of the much publicized findings have faded away. In other words, the report demonstrates that, after a rigorous winnowing process, only a small number of convincing associations have emerged from the masses of individual factors that received prolonged study. There is still much we just do not know. Reports such as this serve a useful function, discriminating what stands up and is important from what has not panned out.

By critically evaluating all of the pertinent evidence on the topic of diet and cancer, the World Cancer Research Fund report helps to put a massive amount of data into a much needed perspective. Similarly, Doll and Peto's

The Causes of Cancer provides an overview that enables us to gauge the relative importance—based on the best available evidence—of different exposures that play a role in the causation of cancer.[4] The fact that *The Causes of Cancer*, which was published in 1981, has not required major revision in the intervening years testifies both to its judicious and comprehensive assessment and to the slow pace of accumulation of new knowledge.

Since the 1980s and increasingly in the 1990s, a number of epidemiologists have expressed discomfort with the tendency of modern epidemiology to focus narrowly on the association of a discrete risk factor, or risk factors, with a specific disease, to the neglect of the social context in which life is lived. This approach, which has been labeled "risk factor epidemiology," dates from the post-World War II period and reflects the success of epidemiology and the development of its methods during the past half century. Rather than denying the value of this approach, critics point out how much it leaves out. The process of reevaluation has been prompted by an awareness of new challenges that have arisen in the past twenty years. These include the reemergence of infectious diseases in the developed world, the global AIDS epidemic, the impact of social inequalities on health, and problems posed by large shifts in populations, both within countries and across national boundaries. An approach that isolates a small number of behavioral factors (smoking, hormone use, alcohol consumption) is clearly not adequate to the task of grappling with the impact of such complex processes on health exerting their effects throughout the life span. Thus, critics have called for a broader, integrative vision of epidemiology—one that can accommodate social, economic, and ecological/environmental realities and their effects on human health, as well as rapidly evolving knowledge of the mechanisms of disease at the molecular level.[5] Such an approach holds out the potential to illuminate many influences acting at the population level and early in life that have profound effects on adult health. Opening up the field of study to include such factors as social class, income, education, occupation, residence, environmental quality, health care, social support, psychosocial and cultural influences, and leisure time activities will likely lead to the identification of important correlations and opportunities for interventions at the population level. In attempting to delineate a different future for epidemiology, these thinkers emphasize that the history of the discipline provides examples of a more contextualized and less disembodied approach.

These stimulating efforts to elaborate a new outlook for epidemiology and public health are closely related to what I have referred to as the "sociology of health hazards." Both take as a central premise that epidemiologists are part of a particular society at a specific point in history. Both take as a given that science is not neutral, and both draw attention to the constraints and biases of the dominant paradigm of a discipline that make it more likely that certain questions will be pursued than others. Finally, both emphasize the importance of social context, which includes the material circumstances of life as well as the beliefs, interests, and outlooks of specific groups.

The decoding of the human genome at the beginning of the new millennium marked the beginning of an era in which the pace of discovery of new genetic factors affecting the risk of many diseases is bound to accelerate. Current thinking suggests that a disease like breast cancer—and other complex chronic diseases—is due to the combined effects of many low-penetrance genes, possibly more than one hundred.[6] Reports of specific genetic variants and their influence on the risk of disease will continue to proliferate, as will reports of the joint effects of genetic and environmental determinants of disease. Given the enormous number of candidate genes and new methods permitting "whole-genome scanning," it is widely recognized that the opportunities for publishing premature findings before their implications are worked out will dwarf anything that we have seen in the past.[7] In this new era, it will be all the more important to maintain a critical stance that attempts to see things in perspective, that distinguishes between flimflam and useable knowledge, and that also keeps in view the social, psychological, and professional realities under which science is conducted and disseminated.

APPENDIX A
List of Interviews

Roger A. Jenkins, Chemist, Oak Ridge National Laboratory, Oak Ridge, Tennessee; currently consultant; October 2, 2003 (e-mail).

Robert K. Adair, Professor Emeritus, Department of Physics, Yale University; October 7, 2003, and May 18, 2004 (in person and e-mail).

Melvin First, Professor of Environmental Health Engineering, Emeritus, Harvard School of Public Health; November 21, 2003 (e-mail).

Marilie D. Gammon, Professor of Epidemiology, University of North Carolina, Chapel Hill; January 20, 2004 (in person).

David A. Savitz, Professor of Epidemiology, University of North Carolina, Chapel Hill; currently Professor of Community and Preventive Medicine and Director, Epidemiology, Biostatistics, and Disease Prevention Institute, Mount Sinai School of Medicine, New York; January 21, 2004 (in person).

Mary S. Wolff, Professor of Community and Preventive Medicine, Mount Sinai School of Medicine, New York; July 7, 2004 (in person).

Steven D. Stellman, Professor of Epidemiology, Mailman School of Public Health, Columbia University, New York; October 13, 2004 (in person).

Regina M. Santella, Professor of Environmental Health Sciences, Mailman School of Public Health, Columbia University, New York; January 20, 2005 (e-mail).

Barbara Balaban, M.S.W., breast cancer activist, January 20, 2005 (telephone).

Ruth Allen, U.S. Environmental Protection Agency, Washington, D. C.; January 28, 2005 (e-mail and telephone).

Gwen Collman, Branch Chief, Genes and Environment Initiative, National Institute of Environmental Health Sciences; February 4, 2005 (telephone).

Fran Visco, Head, National Breast Cancer Coalition, Washington, D. C.; May 10, 2005 (e-mail and telephone).

Dale Sandler, Chief, Chronic Disease Epidemiology Group, National Institute of Environmental Health Sciences; August 5, 2006 (e-mail).

John S. Neuberger, Professor of Epidemiology, University of Kansas, Kansas City; August 16, 2006, and September 11, 2006 (e-mail and telephone).

APPENDIX B
How Findings Can Be Reported in a Way That Puts Them in Perspective

The two examples that follow do an excellent job of qualifying the findings that were reported (one based on a pooling of cohort studies of alcohol consumption and colorectal cancer and the other based on a meta-analysis of studies of hormone replacement therapy and colorectal cancer). Many other examples could have been selected. These two examples of how one can communicate the limitations of a study and attempt to put the findings in perspective for specialists as well as laypeople and reporters contrasts with some of the reporting of results examined in the foregoing chapters. Reporting one's results with the requisite critical detachment makes it difficult for a reader to misunderstand or inflate their significance.

EXAMPLE 1

Eunyoung Cho, et al. "Alcohol Intake and Colorectal Cancer: A Pooled Analysis of 8 Cohort Studies." *Annals of Internal Medicine* 140 (2004), 603–613.

From the abstract:

> Results: Compared with nondrinkers, the pooled multivariate relative risks were 1.16 (95% CI 0.99 to 1.36) for persons who consumed 30 to less than 45 g/d and 1.41 (CI 1.16 to 1.72) for those who consumed 45 g/d or greater.
>
> Limitations: The study included only one measure of alcohol consumption at baseline and could not investigate lifetime alcohol consumption, alcohol consumption at younger ages, or changes in alcohol consumption during follow-up. It could also not examine drinking patterns or duration of alcohol use.
>
> Conclusions: A single determination of alcohol intake correlated with a modest relative elevation in colorectal cancer rate, mainly at the highest levels of alcohol intake.

EXAMPLE 2

Grodstein, F., et al. "Postmenopausal Hormone Therapy and the Risk of Colorectal Cancer: A Review and Meta-Analysis." *American Journal of Medicine* 106 (May 1999), 574–582.

The cover page of this article includes a "Commentary," which makes the following comments clarifying the limitations of the reported results:

"In their review of predominantly observational studies, Grodstein et al. compared women who had ever used postmenopausal hormones with women who had never used them and found a 20% reduction in cancer of the colon and a 19% reduction in cancer of the rectum. . . . This review consolidates the results of 18 studies, only 1 of which is a randomized controlled trial. The authors have identified the design of each study and have weighted the studies in the meta-analysis according to study size. They did not, however, assess the methodological quality of the studies with respect to important criteria such as blinding of data collectors to exposure (i.e. hormone use) or to outcome (i.e. colon or rectal cancer) depending on the study design and completeness of follow up of study participants in the prospective studies. . . . The authors provided limited data on the type of hormone use but note that only 3 studies examined the effect of combined oestrogen and progestin use. For these reasons, validation of the findings of this review by future reviews on the same topic will be important to increase confidence in the results." Dinah Gould, Alba DiCenso.

NOTES

1. INTRODUCTION: Toward a Sociology of Health Hazards in Daily Life

1. P. Slovic, ed., *The Perception of Risk* (London, UK: Earthscan, 2002); C. R. Sunstein, *Risk and Reason: Safety, Law, and the Environment* (Cambridge: Cambridge University Press, 2002); N. Pidgeon, R. E. Kasperson and P. Slovic (eds), *The Social Amplification of Risk* (Cambridge: Cambridge University Press, 2003).

2. In addition to the rise of epidemiology and environmentalism, it is also likely that the countercultural critique of capitalism of the 1960s made possible a novel critique of corporate practices, including those with harmful effects on health and the environment.

3. In the area of cancer, in addition to cigarette smoking and lung cancer, these include heavy alcohol consumption and cancers of the upper respiratory and alimentary tracts, exposure to solar radiation and skin cancer, exposure to estrogen replacement therapy as a cause of endometrial cancer, exposure to ionizing radiation and certain chemicals as causes of specific cancers, as well as the role of specific viruses and bacteria. In addition, epidemiology played a critical role in identifying a variety of risk factors for coronary heart disease. A listing of some of the major achievements of epidemiology is given by Kuller (L. Kuller, "Is Phenomenology the Best Approach to Health Research?" *American Journal of Epidemiology* 166 (2007): 109–1115).

4. G. Kolata, "Environment and Cancer: The Links Are Elusive," *New York Times*, December 13, 2005. This article provides one of the clearest and most insightful discussions of this issue that I have seen.

5. J. P. A. Ioannidis, "Why Most Published Research Findings Are False," *PLoS Medicine* 2 (August 2005): e124.

6. G. Taubes, "Epidemiology Faces Its Limits," *Science* 269 (July 14, 1995): 164–169.

7. J. Niederdeppe and A. G. Levy, "Fatalistic Beliefs About Cancer Prevention and Three Prevention Behaviors," *Cancer Epidemiology Biomarkers and Prevention* 16, 5 (May 1, 2007): 988–1003, doi:10.1158/1055-9965.EPI-06-0608.

8. Of course, "benefits"—whether of pharmaceutical products, diet supplements, or other health products—are also routinely hyped, in what would appear to be the "flip side" of the hyping of dangers.

9. Thomas S. Kuhn, *The Structure of Scientific Revolutions* (2nd ed.) (Chicago: University of Chicago Press, 1970); Paul Feyerabend, *Against Method* (London: Verso, 2002); Robert K. Merton, *The Sociology of Science: Theoretical and Empirical Investigations* (Chicago: University of Chicago Press, 1973); Michel Foucault, *Power/Knowledge: Selected Interviews and Other Writings, 1972–77*, ed. Colin Gordon (New York: Pantheon, 1980).

10. K. M. Cummings, A. Brown, and R. O'Connor, "The Cigarette Controversy," *Cancer Epidemiology Biomarkers and Prevention* (June 2007): 1070–1076; Richard Kluger, *Ashes to Ashes: America's Hundred-Year Cigarette War, the Public Health, and the Unabashed Triumph of Philip Morris* (New York: Random House, 1997); Gerald Markowitz and David Rosner, *Deceit and Denial: The Deadly Politics of Industrial Pollution* (Berkeley: University of California Press, 2002).

11. D. Michaels and C. Monforton, "Manufacturing Uncertainty: Contested Science and the Protection of the Public's Health and Environment," *American Journal of Public Health* Supplement 95 (2005): S39–S48.

12. A. C. Revkin, "Bush vs. the Laureates: How Science Became a Partisan Issue," *New York Times*, October 19, 2004; Chris Mooney, *The Republican War on Science* (New York: Basic Books, 2005); K. M. Rest and M. H. Halpern. "Politics and the Erosion of Federal Scientific Capacity: Restoring Scientific Integrity to Public Health Science," *American Journal of Public Health* 97 (2007): 1939–1944.

13. L. Hardell et al., "Secret Ties to Industry and Conflicting Interests in Cancer Research," *American Journal of Industrial Medicine* 50 (2007): 227–233. M. Newton and A. L. Young. "The Story of 2,4,5-T: A Case Study of Science and Societal Concerns," *Environmental Science and Pollution Research* 11 (2004): 207–208; P. Cole et al., "Dioxin and Cancer: A Critical Review," *Regulatory Toxicology and Pharmacology* 38 (2003): 378–388; J. D. Boice Jr. and J. K. McLaughlin, "Epidemiologic Studies of Cellular Telephones and Cancer Risk: A Review," SSI Report (Statens Strålskyddsinstitut) 16 (2002); R. L. Barnes, S. K. Hammond, and S. A. Glantz, "The Tobacco Industry's Role in the 16 Cities Study of Secondhand Tobacco Smoke: Do the Data Support the Stated Conclusions?" *Environmental Health Perspectives* 114 (2006): 1890–1897; C. V. Phillips, "Warning: Anti-Tobacco Activism May Be Hazardous to Epidemiologic Science," *Epidemiologic Perspectives and Innovations* 4 (2007), doi:10.1186/1742-5573-4-13.

14. To give just a few examples restricted to the United States, the auto industry has effectively stymied progress on pollution control; the mining industry has almost entirely done away with health and safety enforcement; the food industry has brought about tremendous limitations in agriculture inspection, especially of meat; and the Food and Drug Administration is "so overwhelmed by the flood of imports that it is incapable of protecting the public from unsafe drugs, medical devices and food" from abroad, and particularly from China (G. Harris, "For F.D.A., A Major Backlog Overseas," *New York Times*, January 29, 2008.

15. J. D. Graham, K. Clemente, and R. Glass, "Breast Cancer: What Are the Perceived Risk Factors?" *Harvard Center for Risk Analysis* 4 (1996): 1–2.

16. Slovic, *The Perception of Risk*, 264–274.

17. Ibid.

18. Ibid.

19. J. M. Last, *A Dictionary of Epidemiology* (4th ed.) (New York: Oxford University Press, 2001); P. Boffetta et al. "'Environment' in Cancer Causation and Etiological Fraction: Limitations and Ambiguities," *Carcinogenesis* 28 (May 2007): 913–915.

20. H. Harris and A. O'Connor, "On Autism's Cause, It's Parents vs. Research," *New York Times*, June 25, 2005.

21. The power of these movies stems in no small part from the deep-seated fear of a deadly toxin seeping into our environment and imperceptibly causing disease. Even though the science often cannot provide clear-cut evidence of cause and effect in such situations, there is a larger truth behind these stories and the associated fear, and this is that there is always the potential for the gradual accumulation of toxins that could lead to disease. And some disease clusters due to a "point source" of pollution undoubtedly go undetected.

22. Phillips, "Warning: Anti-Tobacco Activism"; D. T. Levy, E. A. Mumford, K. M. Cummings, E. A. Gilpin, G. Giovino, A. Hyland, D. Sweanor, and K. E. Warner. "The Relative Risks of a Low-Nitrosamine Smokeless Tobacco Product Compared with Smoking Cigarettes: Estimates of a Panel of Experts," *Cancer Epidemiology Biomarkers and Prevention* 13 (2004): 2035–2042.

23. Kolata, "Environment and Cancer."

24. B. N. Ames and L. S. Gold. "Paracelsus to Parascience: The Environment and Cancer Distraction," *Mutation Research* 447 (2000): 3–13.

25. M. L. Wald, "With New Data, A Debate on Low-Level Radiation," *New York Times*, July 7, 2005; G. Kolata, "For Radiation, How Much Is Too Much?" *New York Times*, November 27, 2001.

26. E. L. Wynder, I. T. T. Higgins, and R. E. Harris, "The Wish Bias," *Journal of Clinical Epidemiology* 43 (1990): 619–621; Ioannidis, "Why Most Published Research Findings Are False," 2005.

27. Cited in Taubes, "Epidemiology Faces Its Limits."

28. To give just one example of what I have in mind, it has been widely accepted that roughly 50,000 deaths from heart disease a year are due to exposure to secondhand tobacco smoke. Review articles and agency reports that attempt to make a case for secondhand smoke being an important cause of heart disease are unlikely to mention, no less to give weight to, certain highly relevant facts. First, active smoking is one of a number of risk factors for heart disease and is associated with less than a doubling of the risk. Second, exposure to secondhand smoke is orders of magnitude more dilute that the smoke the active smoker inhales. Third, since the 1950s the death rate from heart disease has declined dramatically in both smokers and nonsmokers, by about 50–60 percent. These would seem to be pertinent facts to mention when considering a possible effect of secondhand smoke exposure on heart disease risk. However, giving these facts due consideration would have the effect of reducing the importance of the factor under study.

29. The Oxford epidemiologist Sir Richard Peto is quoted as saying, "Epidemiology is so beautiful and provides such an important perspective on human life and death, but an incredible amount of rubbish is published." G. Taubes, "Unhealthy Science: Why Can't We Trust Much of What We Hear About Diet, Health, and Behavior-Related Diseases?" *New York Times Magazine*, September 16, 2007, 52–80.

30. S. Shane, "Debating the Evidence of Gulf War Illnesses," *New York Times*, November 16, 2004.

31. L. A. Cole, *Element of Risk: The Politics of Radon* (New York: Oxford University Press, 1993), 101.

32. Ibid., 87–88.

33. Grady, D. "Pregnancy Problems Tied to Caffeine." *New York Times*, January 21, 2008.

34. M. Douglas and A. Wildavsky, *Risk and Culture: An Essay on the Selection of Technological and Environmental Dangers* (Berkeley: University of California Press, 1982).

35. B. N. Ames, M. Profet, and L. S. Gold, "Dietary Pesticides (99.99% All Natural)," *Proceedings of the National Academy of Sciences USA* 87 (1990): 7777–7781.

36. D. Kahneman and A. Tversky, "Judgment Under Uncertainty: Heuristics and Biases," *Science* 185 (September 1974): 1124–1131.

37. D. Ropeik and G. Gray, *Risk! A Practical Guide for Deciding What's Really Safe and What's Really Dangerous in the World Around You* (Boston: Houghton Mifflin, 2002).

2. EPIDEMIOLOGY: Its Uses, Strengths, and Limitations

1. The many achievements of epidemiology need publicizing because epidemiology has come in for a good deal of criticism (G. Taubes, "Epidemiology Faces Its Limits," *Science* 269 (July 14, 1995), 164–169; G. Taubes, "Unhealthy Science: Why Can't We Trust Much of What We Hear About Diet, Health, and Behavior-Related Diseases?" *New York Times Magazine* (September 16, 2007)). While Taubes makes some valid and important points, his assessment of epidemiology is, I believe, misinformed.

2. K. J. Rothman, *Epidemiology: An Introduction* (New York: Oxford University Press, 2002).

3. For further reading, I strongly recommend Leon Gordis' textbook *Epidemiology* (3rd ed.), Philadelphia: Saunders, 2004. This comes as close to a beginner-friendly yet comprehensive introduction to the field as one can find.

4. K. M. Flegal et al. "Cause-Specific Excess Deaths Associated with Underweight, Overweight, and Obesity," *Journal of the American Medical Association* 298 (November 7, 2007), 2028–2037.

5. K. A. Matthews et al., "Prior Use of Estrogen Replacement Therapy, Are Users Healthier Than Nonusers?" *American Journal of Epidemiology* 143 (May 15, 1996): 971–978; C. A. Derby et al., "Correlates of Postmenopausal Estrogen Use and Trends Through the 1980s in Two Southeastern New England Communities," *American Journal of Epidemiology* 137 (1993): 1125–1135; C. B. Johannes et al., "Longitudinal Patterns and Correlates of Hormone Replacement Therapy Use in Middle-Aged Women," *American Journal of Epidemiology* 140 (1994): 439–452.

6. Unlike the situation of smoking, where we have people who never smoked as the unexposed group, when studying diet, for example, there is no unexposed group, so we can only examine contrasts between different degrees of exposure to a given dietary factor.

7. R. P. Beasley, "Hepatitis B Virus: The Major Etiology of Hepatocellular Carcinoma," *Cancer* 61 (1988): 1942–1956.

8. N. Muñoz et al., "HPV in the Etiology of Human Cancer," *Vaccine* 24, S3 (2006), S1–S10.

9. Beasley, "Hepatitis B Virus."

10. J. E. Enstrom and G. C. Kabat, "Environmental Tobacco Smoke and Tobacco Related Mortality in a Prospective Study of Californians, 1960-98," *BMJ* 326 (May 17, 2003), 1057–1066.

11. International Agency for Research on Cancer (IARC), *IARC Monograph 44: Alcohol Drinking* (Lyon, France: IARC, 1988), 169.

12. G. A. Colditz et al., "Family history, age, and risk of breast cancer: prospective data from the Nurses' Health Study," *Journal of the American Medical Association* 270 (1993): 338–343.

13. C. W. Bain et al., "Early Age at First Birth and Decreased Risk of Breast Cancer," *American Journal of Epidemiology* 114 (1981): 705–709.

14. Enstrom and Kabat, "Environmental Tobacco Smoke."

15. N. R. Shah, J. Borenstein, and R. W. Dubois, "Postmenopausal Hormone Therapy and Breast Cancer: A Systematic Review and Meta-analysis," *Menopause* 12 (2005): 653–655.

16. Z. A. Stein, "Silicone Breast Implants: Epidemiological Evidence of Sequelae," *American Journal of Public Health* 89 (April 1999), 484–487.

17. A. K. Hackshaw, M. R. Law, and N. J. Wald, "The Accumulated Evidence on Lung Cancer and Environmental Tobacco Smoke," *BMJ* 315 (1997): 980–988.

18. G. Pershagen et al., "Residential Radon Exposure and Lung Cancer in Sweden," *New England Journal of Medicine* 330 (1994): 159–164.

19. S. Greenland et al., "A Pooled Analysis of Magnetic Fields, Wire Codes, and Childhood Leukemia," *Epidemiology* 12 (2001): 472–474.

20. C. A. Pope III et al., "Particulate Air Pollution as a Predictor of Mortality in a Prospective Study of U.S. Adults," *American Journal of Respiratory and Critical Care Medicine* 151 (1995), 669–674.

21. F. Laden et al., "Plasma Organochlorine Levels and the Risk of Breast Cancer: An Extended Follow-Up in the Nurses' Health Study," *International Journal of Cancer* 91 (2001): 568–574.

22. Stein, "Silicone Breast Implants."

23. U.S. Environmental Protection Agency, *Respiratory Health Effects of Passive Smoking: Lung Cancer and Other Disorders* (Washington, D. C.: U.S. Department of Health and Human Services, 1992); K. Steenland, "Passive Smoking and the Risk of Heart Disease," *Journal of the American Medical Association* 267 (January 1992), 94–99; National Research Council, *Health Effects of Exposure to Radon: BEIR VI* (Washington, D. C.: National Academy Press, 1999).

24. K. J. Rothman and S. Greenland, "Causation and Causal Inference in Epidemiology," *American Journal of Public Health* (supplement) 95, S1 (2005): S144–S150.

25. Ibid.

26. Ibid.

27. C. V. Phillips and K. J. Goodman, "The Missed Lessons of Sir Austin Bradford Hill," *Epidemiologic Perspectives & Innovations* 1 (October 2004): 3, doi:10.1186/1742-5573-1-3.

28. A. B. Hill, "The Environment and Disease: Association or Causation?" *Proceedings of the Royal Society of Medicine* 58 (1965): 295–300.

29. U.S. Public Health Service. *Smoking and Health: Report of the Advisory Committee to the Surgeon General of the Public Health Service*, PHS Publication 1103,

U.S. Department of Health, Education, and Welfare, Public Health Service, Center for Disease Control, 1964.

30. A. Morabia, "On the Origins of Hill's Causal Criteria," *Epidemiology* 2 (1991): 367–369.

31. R. M. Lucas and A. J. McMichael, "Association or Causation: Evaluating Links Between 'Environment and Disease,'" *Bulletin of the World Health Organization* 83 (2005): 792–795.

32. M. Susser, *Causal Thinking in the Health Sciences: Concepts and Strategies of Epidemiology* (New York: Oxford University Press, 1973).

33. Morabia, "On the Origins of Hill's Causal Criteria."

34. Enstrom and Kabat, "Environmental Tobacco Smoke."

35. J. M. Elwood, et al., "Alcohol, Smoking, and Social and Occupational Factors in the Aetiology of Cancer of the Oral Cavity, Pharynx and Larynx," *International Journal of Cancer* 34 (1984): 603–612.

36. D. Grady et al., "Hormone Replacement Therapy and Endometrial Cancer Risk: A Meta-analysis," *Obstetrics and Gynecology* 85 (1995): 304–313.

37. Enstrom and Kabat, "Environmental Tobacco Smoke."

38. G. A. Colditz, "Relationship Between Estrogen Levels, Use of Hormone Replacement Therapy, and Breast Cancer (Review)," *Journal of the National Cancer Institute* 90 (1998): 814–823.

39. M. Egger, M. Schneider, and G. Davey Smith, "Spurious Precision? Meta-analysis of Observational Studies," *BMJ* 316 (1998): 140–144.

40. S. Shapiro, "Meta-analysis/Shmeta-analysis," *American Journal of Epidemiology* 140 (1994): 771–778.

41. Egger et al., "Spurious Precision? Meta-analysis of Observational Studies."

42. Shapiro, "Meta-analysis/Shmeta-analysis."

43. A. Morabia, "Risky Concepts: Methods in Cancer Research," *American Journal of Public Health* 91 (March 2001): 355–357.

44. Ibid.

45. L. M. Schwartz et al., "Ratio Measures in Leading Medical Journals: Structured Review of Accessibility of Underlying Absolute Risks," *BMJ* 333 (October 2006), 1248, doi:10.1136/bmj.38985.564317.7C.

46. D. Sanghavi, "Treat Me: The Crucial Health Stat You've Never Heard of," *Slate*, September 26, 2006, at http://www.slate.com/toolbar.aspx?action=print&id=2150354.

47. Schwartz et al., "Ratio Measures in Leading Medical Journals."

48. Morabia, "Risky Concepts."

49. M. Marmot, *The Status Syndrome: How Social Standing Affects Our Health and Longevity* (London: Bloomsbury, 2005).

3. DOES THE ENVIRONMENT CAUSE BREAST CANCER?

1. Barbara Balaban, interview with author, January 20, 2005.

2. T. R. Holford et al., "Changing Patterns in Breast Cancer Incidence Trends," *Journal of the National Cancer Institute Monographs* No. 36 (2007): 19–25.

3. P. M. Ravdin et al., "The Decrease in Breast-Cancer Incidence in 2003 in the United States," *New England Journal of Medicine* 356 (April 19, 2007): 1670–1674.

4. The most likely source of the 30 percent figure is a reported 27 percent higher breast cancer *mortality* rate in Nassau and Suffolk compared to the United States rate based on data from the National Center for Health Statistics. This figure was routinely cited as incidence, rather than mortality. Susan Jenks, "Researchers to Comb Long Island for Potential Cancer Factors," *Journal of the National Cancer Institute* 86 (January 19, 1994): 88–89.

5. G. Kolata, "The Epidemic That Wasn't," *New York Times*, August 29, 2002.

6. Regina Santella, interview with author, January 20, 2005.

7. J. D. Graham, K. Clemente, and R. Glass, "Breast Cancer: What Are the Perceived Risk Factors?" *Harvard Center for Risk Analysis* 4 (1996), 1–2.

8. Marilie Gammon, interview with author, January 20, 2004; follow-up interview, January 19, 2005.

9. S. R. Sturgeon et al., "Geographic Variation in Mortality from Breast Cancer Among White Women in the United States," *Journal of the National Cancer Institute* 87 (December 20, 1995): 1846–1853; M. Kulldorff et al., "Breast Cancer Clusters in the Northeast United States: A Geographic Analysis," *American Journal of Epidemioliology* 146 (July 15, 1997): 161–170.

10. For a number of reasons, a higher number of people with a history of a disease like breast cancer in a particular community may not be evidence of any shared common environmental exposure. Apparent cancer clusters are more likely to have a common underlying cause if the cancer involved is rare and if it manifests itself in a group that does not usually develop this cancer. Neither of these conditions held for breast cancer on Long Island. Most investigations of cancer clusters turn out not to have an identifiable common cause. See Atul Gawande, "The Cancer Cluster Myth," *The New Yorker*, February 8, 1999.

11. Balaban, interview with author.

12. Davis' claims about increased cancer rates due to environmental pollution came in for harsh criticism from eminent epidemiologists, such as Richard Doll and Richard Peto. See Karen Wright, "Going by the Numbers," *New York Times Magazine*, December 15, 1991.

13. Mary Wolff, interview with author, July 20, 2004.

14. Interview with Gwen Collman, February 4, 2005; National Cancer Institute Web site, http://www.epi.grants.cancer.gov.

15. National Cancer Institute, "Study of Elevated Breast Cancer Rates in Long Island," http://www.epi.grants.cancer.gov/LIBCSP/PublicLaw.html.

16. Dr. Samuel Broder, who was director of the National Cancer Institute at time, commented that, "I don't think it's a good idea for the Congress, in general terms, to identify a scientific problem and then issue detailed instructions to an institute." Brodeur identified the rush to enact the legislation and the lack of give-and-take between the politicians and the cancer institute as having seriously compromised the study. "It's not an optimal situation when the Congress doesn't allow time for dialogue, or doesn't even want dialogue," he said. "If you have dialogue at the beginning, you can come to an understanding about the scope, the limits and the expectations of a project." Dan Fagin, "Tattered

Hopes: A $30-million federal study of breast cancer and pollution on LI has disappointed activists and scientists." *Newsday*, July 28, 2002.

17. D. Fagin, "So Many Things Went Wrong: Costly Search for Links Between Pollution and Breast Cancer Was Hobbled from the Start," *Newsday*, July 29, 2002.

18. In an interview with *Newsday* in 2000, Dr. Ellen Heineman, the National Cancer Institute official overseeing the GIS project, gave a more realistic appraisal of the utility of the GIS for identifying the causes of breast cancer: "A 'beginning' is also how Ellen Heineman . . . described the system. She emphasized that the data will allow researchers to test their hypotheses and spark further research, but it won't lead them to definite conclusions. 'It's a starting point, a jumping board. People who are putting all their hopes in this system will be terribly disappointed,' she told the audience." Heather Sokoloff, Cancer map of LI unveiled. *Newsday*, September 21, 2000.

19. T. Coborn, D. Dumanoski, and J. P. Myers, *Our Stolen Future: Are We Threatening Our Fertility, Intelligence, and Survival?* (New York: Plume Books, 1996).

20. H.-O. Adami et al., "Organochlorine Compounds and Estrogen-Related Cancers in Women," *Cancer Causes and Control* 6 (1995): 551–566; U. G. Ahlborg et al., "Organochlorine Compounds in Relation to Breast Cancer, Endometrial Cancer, and Endometriosis: An Assessment of the Biological and Epidemiological Evidence," *Critical Reviews in Toxicology* 25 (1995): 463–531.

21. T. Key and G. Reeves, "Organochlorines in the Environment and Breast Cancer," *BMJ* 308 (June 11, 1994): 1520–1521.

22. S. H. Safe, "Is There an Association Between Exposure to Environmental Estrogens and Breast Cancer?" *Environmental Health Perspectives* 105, Supplement 3 (April 1997): 675–678.

23. D. H. Phillips, "DNA Adducts as Markers of Exposure and Risk," *Mutation Research* 577 (2005): 284–292.

24. S. Jenks, "Researchers to Comb Long Island for Potential Cancer Factors," *Journal of the National Cancer Institute* 86 (1994): 88–89.

25. R. Ochs, "The LI Breast Cancer Study," *Newsday*, December 10, 1996.

26. Ibid.

27. M. S. Wolff et al., "Blood Levels of Organochlorine Residues and Risk of Breast Cancer," *Journal of the National Cancer Institute* 85 (1993): 648–652.

28. Ibid.

29. D. J. Hunter and K. T. Kelsey, "Pesticide Residues and Breast Cancer: The Harvest of a Silent Spring?" *Journal of the National Cancer Institute* 85 (April 21, 1993): 598–599.

30. E. E. Calle, et al., "Organochlorines and Breast Cancer Risk," *CA A Cancer Journal for Clinicians* 52 (2002): 301–309.

31. Ibid.

32. M. Lopez-Cervantes et al., "Dichlorodiphenyldichloroethane Burden and Breast Cancer Risk: A Meta-analysis of the Epidemiologic Evidence," *Environmental Health Perspectives* 112 (2004): 207–214.

33. M. S. Wolff et al., "Risk of Breast Cancer and Organochlorine Exposure," *Cancer Epidemiology Biomarkers Prevention* 9 (2000): 271–277.

34. M. D. Gammon et al., "Environmental Toxins and Breast Cancer on Long Island. II. Organochlorine Compound Levels in Blood," *Cancer Epidemiology Biomarkers & Prevention* 11 (2002): 686–697.

35. M. D. Gammon and R. M. Santella, "Reply to F. Perrera," *Cancer Epidemiology Biomarkers & Prevention* 12 (2003): 75–76.

36. B. MacMahon et al., "Coffee and Cancer of the Pancreas," *New England Journal of Medicine* 304 (1981): 630–633.

37. V. L. Ernster et al., "Effects of Caffeine-Free Diet on Benign Breast Disease: A Randomized Trial," *Surgery* 91 (1982), 263–267.

38. W. C. Willett, "Prospective Studies of Diet and Breast Cancer," *Cancer* 74, 3 Supplement (1994): 1085–1089.

39. In response to the advocates' desire to see a wider range of compounds included in the study, Gammon and colleagues drafted a report tabulating available evidence for thirty-six additional compounds, including both evidence of carcinogenicity in laboratory experiments and the availability of an adequate biological marker. "LIBCSP Year 2 Progress Report: Other Chemicals of Community Interest."

40. D. Fagin, "Tattered Hopes; A $30-Million Dollar Federal Study of Breast Cancer and Pollution on LI Has Disappointed Activists and Scientists," *Newsday*, July 28, 2002; "So Many Things Went Wrong: Costly Search for Links Between Pollution and Breast Cancer Was Hobbled from the Start, Critics Say," *Newsday*, July 29, 2002; "Still Searching: A Computer Mapping System Was Supposed to Help Unearth Information About Breast Cancer and the Environment," *Newsday*, July 30, 2002.

41. Kolata, "The Epidemic That Wasn't."

42. Ibid.

43. Ibid.

44. Marilie Gammon, interview with author, January 20, 2004.

45. Ibid.

46. Ibid.

47. Steven Stellman, interview with author, October 13, 2004.

48. Institute of Medicine. *Committee to Review the Health Effects in Vietnam Veterans of Exposure to Herbicides. Veterans and Agent Orange: Health Effects of Herbicides Used in Vietnam.* (Washington, D.C.: National Academy of Sciences Press, 1994), 283.

49. Ibid.

50. Mary Wolff, interview with author, July 7, 2004.

51. G. Kolata, "Reversing Trend, Big Drop Is Seen in Breast Cancer," *New York Times*, December 15, 2006; P. M. Ravdin et al., "The Decrease in Breast-Cancer Incidence," 2007.

52. C. McNeil, "Breast Cancer Decline Mirrors Fall in Hormone Use, Spurs Both Debate and Research," *Journal of the National Cancer Institute* 99 (February 21, 2007): 266–267.

53. V. C. Jordan, "SERMs: Meeting the Promise of Multifunctional Medicines," *Journal of the National Cancer Institute* 99 (March 7, 2007): 350–356.

54. G. Kolata, "Hormones and Cancer: Assessing the Risks," *New York Times*, December 26, 2006.

55. Ibid.

4. ELECTROMAGNETIC FIELDS: The Rise and Fall of a "Pervasive Threat"

1. L. I. Kheifets et al., "Electric and Magnetic Fields and Cancer: Case Study," *American Journal of Epidemiology* 154 (supplement) (2000): S50–S59.

2. D. A. Savitz and A. Ahlbom, "Electromagnetic Fields and Radiofrequency Radiation, in *Cancer Epidemiology*, ed. D. Schottenfeld and J. R. Frameni Jr. (Oxford: Oxford University Press, 2006).

3. D. A. Bromley, *The President's Scientist: Reminiscences of a White House Science Advisor* (New Haven: Yale University Press, 1994); R. L. Park, *Voodoo Science: The Road from Foolishness to Fraud* (New York: Oxford University Press, 2000).

4. Wertheimer and Leeper, "Electrical Wiring Configurations," 283.

5. L. Tomenius, "50-Hz Electromagnetic Environment and the Incidence of Childhood Tumors in Stockholm County," *Biolectromagnetics* 7 (1986): 191–207.

6. For a review of the early occupational studies, see D. A. Savitz and E. E. Calle, "Leukemia and Occupational Exposure to Electromagnetic Fields: Review of Epidemiologic Surveys," *Journal of Occupational Medicine* 29 (1987): 47–51.

7. David Savitz, interview with author, January 21, 2004.

8. Savitz et al., "Case-Control Study of Childhood Cancer."

9. D. A. Savitz, N. E. Pearce, and C. Poole, "Methodological Issues in the Epidemiology of Electromagnetic Fields and Cancer," *Epidemiologic Reviews* 11 (1989): 59–78.

10. Ibid., 61.

11. Ibid., 75.

12. C. Poole and D. Trichopoulos, "Extremely Low-Frequency Electric and Magnetic Fields and Cancer," *Cancer Causes and Control* 2 (1991): 267–276.

13. Ibid., 273.

14. ORAU Panel on Health Effects of Low-Frequency Electric and Magnetic Fields, "EMF and Cancer (Letter)," *Science* 260 (April 2, 1993): 14–16.

15. Ibid., 15

16. P. Brodeur, *Currents of Death: Power Lines, Computer Terminals, and the Attempt to Cover Up Their Threat to Your Health* (New York: Simon and Shuster, 1989).

17. E. R. Adair, "Currents of Death" rectified. A paper commissioned by the IEEE-USA Committee on Man and Radiation in Response to the book by Paul Brodeur (New York: IEEE-USA, 1991).

18. R. G. Stevens, "Electric Power Use and Breast Cancer: A Hypothesis," *American Journal of Epidemiology* 125 (1987): 556–561; G. C. Brainard, R. Kavet, and L. I. Kheifets, "The Relationship Between Electromagnetic Field and Light Exposures to Melatonin and Breast Cancer Risk: A Review of the Relevant Literature," *Journal of Pineal Research* 26 (1999): 65–100.

19. R. K. Adair, "Constraints on Biological Effects of Weak Extremely-Low-Frequency Electromagnetic Fields," *Physical Review A* 43 (1991): 1039–1048. This first paper by Adair on electromagnetic fields built to some extent on a highly influential paper by two physicists that had appeared in 1990: J. C. Weaver and

R. D. Astumian, "Response of Cells to Very Weak Electric Fields," *Science* 247 (1990): 459–462.

20. Robert K. Adair, email message to author, November 22, 2004.

21. R. K. Adair, "Fear of Weak Electromagnetic Fields," *Scientific Review of Alternative Medicine* 3 (1999): 22–23.

22. R. K. Adair, "Static and Low-Frequency Magnetic Fields: Health Risks and Therapies," *Reports on Progress in Physics* 63 (2000): 415–454. Adair had this to say about experimental research on ELF-EMF: "There is strong evidence that fields between 1–10 mV m^{-1} affect biology. However, after a quarter century of intense experimentation and more than 200 results that report biological effects of fields smaller than 50 T at 50-60 Hz, fields that generate induced currents less than 1 mA m^{-1}, no persuasive result showing effects of such weak fields has been forthcoming.

"Is the rejection of so large a set of results—albeit none that are definitive—quite unusual? No! Other areas of science have experienced the phenomenon of having large sets of invalid results purporting to establish pathological science. Recently, there have been several hundred reports of experiments that demonstrated 'cold fusion'; but there is no cold fusion." (p. 437).

23. U.S. Department of Health and Human Services, "Findings of Scientific Misconduct," *NIH Guide* (June 18, 1999).

24. R. K. Adair, "Measurements Described in a Paper by Blackman, Blanchard, Bename, and House Are Statistically Invalid," *Bioelectromagnetics* 17 (1996): 510–511.

25. W. R. Bennett Jr., "Cancer and Power Lines," *Physics Today* (April 1994): 23–29.

26. R. K. Adair, "Biological Responses to Weak 60-Hz Electric and Magnetic Fields Must Vary as the Square of the Field Strength," *Proceedings of the National Academy of Sciences USA* 91 (1994): 9422–9425.

27. National Research Council, *Possible Health Effects of Exposure to Residential Electric and Magnetic Fields* (Washington, D. C.: National Academy Press, 1997).

28. Ibid., 1–2.

29. The makeup of a committee can have a decisive effect on the tenor of its conclusions. In this regard, the background to the formation of the National Research Council's EMF committee is worth noting. According to Adair, the original chairman-designate was David Savitz, who by then was chairman of the epidemiology department at the University of North Carolina. Owing to complaints from Adair and other physicists, first to Frank Press, president of the National Academy of Sciences, and then to Bruce Alberts, his successor, Stevens was appointed as chairman, and Savitz became deputy chairman. In addition, Adair was concerned that "about 40 percent" of the committee members had a bias in favor of health effects from EMF due to their own work. To ensure what he saw as balance, he succeeded in persuading Alberts to add the physicist Richard Garwin to the committee.

30. S. J. London et al., "Exposure to Residential Electric and Magnetic Fields and Risk of Childhood Leukemia," *American Journal of Epidemiology* 134 (1991): 923–937.

31. M. Feychting and A. Ahlbom, "Magnetic Fields and Cancer in Children Residing near Swedish High-Voltage Power Lines," *American Journal of Epidemiology* 138 (1993): 467–481.

32. R. Wilson and A. Shlyakhter, "Re: Magnetic Fields and Cancer in Children Residing near Swedish High-Voltage Power Lines," *American Journal of Epidemiology* 141 (1995): 378–379.

33. M. S. Linet et al., "Residential Exposure to Magnetic Fields and Acute Lymphoblastic Leukemia in Children," *New England Journal of Medicine* 337 (1997): 1–7.

34. National Institute of Environmental Health Sciences, *NIEHS Report on Health Effects from Exposure to Power-Line Frequency Electric and Magnetic Fields*, NIH Publication 99-4493, Washington D.C., 1999.

35. Ibid., 14.

36. Clearly outraged by the biases operating in the NIEHS working group, Adair drafted a ten-page, detailed inventory of biases and errors in the report ranging from the composition of the committee to the interpretation of specific experiments. This was submitted to NIEHS during the public comment period on the document. (R. K. Adair, "A Critique of the NIEHS Working Group Report on Assessment of Health Effects from Exposure to Power-Line Frequency Electric and Magnetic Fields," September 17, 1998).

37. M. B. Bracken et al., "Correlates of Residential Wiring Code Used in Studies of Health Effects of Residential Electromagnetic Fields," *American Journal of Epidemiology* 148 (1998): 467–474; E. Hatch et al., "Do Confounding or Selection Factors of Residential Wiring Codes and Magnetic Fields Distort Findings of Electromagnetic Fields Studies?" *Epidemiology* 11 (2000): 189–198.

38. S. Greenland et al., "A Pooled Analysis of Magnetic Fields, Wire Codes, and Childhood Leukemia," *Epidemiology* 11 (2000): 624–634.

39. A. Ahlbom et al., "A Pooled Analysis of Magnetic Fields and Childhood Leukemia," *British Journal of Cancer* 83 (2000): 692–698.

40. Greenland et al., "A Pooled Analysis of Magnetic Fields, Wire Codes," 632.

41. Ahlbom et al., "A Pooled Analysis of Magnetic Fields and Childhood Leukemia," 697–698.

42. S. Davis, D. K. Mirick, and R. G. Stevens, "Residential Magnetic Fields and the Risk of Breast Cancer," *American Journal of Epidemiology* 155 (2002): 446–454; E. R. Schoenfeld et al., "Electromagnetic Fields and Breast Cancer on Long Island: A Case-Control Study," *American Journal of Epidemiology* 158 (2003): 47–58; S. J. London et al., "Residential Magnetic Field Exposure and Breast Cancer Risk: A Nested Case-Control Study from a Multiethnic Cohort in Los Angeles County, California," *American Journal of Epidemiology* 15 (2003): 158: 969–980.

43. The NIEHS final report issued in 1999 concluded that the accumulated evidence provided "little support that exposure to ELF-EMF is altering melatonin levels in humans" (p. 19).

44. D. P. Sandler, "On Blankets and Breast Cancer (Editorial)," *Epidemiology* 14 (2003): 509.

45. E. W. Campion, "Power Lines, Cancer, and Fear (Editorial)," *New England Journal of Medicine* 337 (1997): 44–46.

46. D. A. Savitz, "Health Effects of Electric and Magnetic Fields: Are We Done Yet?" *Epidemiology* 14 (2003): 15–17.

47. M. Feychting and A. Ahlbom, "Childhood Leukemia and Residential Exposure to Weak Extremely Low Frequency Magnetic Fields," *Environmental Health Perspectives*, Supplement 2 (1995): 59–62.

48. A. B. Hill, "The Environment and Disease: Association or Causation?" *Proceedings of the Royal Society of Medicine* 58 (1965): 295–300.

49. Kheifets, "Electric and Magnetic Fields and Cancer."

5. THE SCIENCE AND POLITICS OF RESIDENTIAL RADON

1. J. H. Lubin and J. D. Boice Jr., "Lung Cancer Risk From Residential Radon: Meta-analysis of Eight Epidemiologic Studies," *Journal of the National Cancer Institute* 89 (January 1, 1997): 49–57.

2. L. Alderson, "A Creeping Suspicion About Radon," *Environmental Health* Perspectives 102 (October 1994): 826–831.

3. National Research Council. *Health Effects of Exposure to Radon.* BEIR VI. (Washington, D. C.: National Academy Press, 1999).

4. National Research Council. *Health Risks of Radon and Other Internally-Deposited Alpha-Emitters.* BEIR IV (Washington, D.C.: National Academy Press, 1988).

5. National Research Council, *Health Effects of Exposure*, 20.

6. National Research Council, *Health Risks of Radon*, 445.

7. Ibid., 446–447.

8. Ibid., 448.

9. L. A. Cole, *Element of Risk: The Politics of Radon* (New York: Oxford University Press, 1993), 11.

10. Cole, *Element of Risk*, 11; R. N. Proctor, *Cancer Wars: How Politics Shapes What We Know and Don't Know About Cancer* (New York: Basic Books, 1995), 216.

11. Proctor, *Cancer Wars*, 203–204.

12. Ibid., 204.

13. J. D. Boice Jr., "Ionizing Radiation," in *Cancer Epidemiology and Prevention*, eds. D. Schottenfeld and J. F. Fraumeni Jr. (Oxford: Oxford University Press, 2006), 259–293.

14. P. Shabecoff, "Radioactive Gas in Soil Raises Concern in Three-State Area," *New York Times*, May 19, 1985.

15. Cole, *Element of Risk*, 12–13. According to Anthony Nero, a physicist at Lawrence Berkeley Laboratory, as quoted in the *New York Times*, "It is fairly clear that the real incentive for activity was that a single area in eastern Pennsylvania was found to contain homes with radon levels that were truly astounding. It finally shook local and national regulators into taking regulatory action. Before they had been only doing research and thinking about the problem." P. Shabecoff, "Issue of Radon: New Focus on Ecology," *New York Times*, September 10, 1986.

16. P. Shabecoff, "Drive to Locate Risk Areas for Radioactive Gas Urged," *New York Times*, May 24, 1985; P. Shabecoff, "E.P.A. Proposes 5-Year Program Aimed at Radioactive Radon Gas," *New York Times*, October 10, 1985.

17. R. Pear, "Safety Standard is Set on Radon in U.S. Homes," *New York Times*, August 15, 1986; "U.S. Says Radon Gas Is States' Fight," *New York Times*, August 16, 1986.

18. U.S. Environmental Protection Agency. Technical support document for the 1992 *Citizen's Guide to Radon*, EPA 400-R-92-011 (Washington, D.C.: U.S. Environmental Protection Agency, 1992).

19. *Atlanta Constitution*, July 15, 1985.

20. *Newsweek*, August 18, 1986, 60–61.

21. P. Shabecoff, "Radioactive Gas in Soil Raises Concern in Three-State Area," *New York Times*, May 19, 1985.

22. K. E. Warner, D. D. Mendez, and P. N. Courant. "Toward a More Realistic Appraisal of the Lung Cancer Risk from Radon: The Effects of Residential Mobility," *American Journal of Public Health* 86 (September 1, 1996): 1222–1227.

23. Cole, *Element of Risk*, 1.

24. Ibid., 150. Cole asks why media attention to radon declined after 1987, when the EPA and the U.S. Surgeon General's office ratcheted up their message in the late eighties. A partial explanation, he argues, is that radon became just another in a long succession of trumpeted health hazards and that saturation set in. But he also proposes that EPA's extremist radon policy coupled with problems of obtaining accurate testing results and effective remediation, as well as the bungled handling of landfill contamination in places like Montclair, New Jersey, undermined the public's confidence in the agency's policy (Cole, 151–152).

25. Ibid., 62, 93–97.

26. Ibid., 82.

27. Ibid., 17–19, 82; Proctor, *Cancer Wars*, 214.

28. Cole, *Element of Risk*, 93; S. S. Epstein, "A Straw Man" (letter), *New York Times*, October 22, 1988; Proctor, *Cancer Wars*, 199, 211; "Cancer Risk from Domestic Radon," *Lancet* 1 (January 14, 1989): 93.

29. U.S. Environmental Protection Agency. *A Citizen's Guide to Radon: What It Is and What to Do About It* (Washington, D.C.: U.S. Environmental Protection Agency, August 1986).

30. Ibid.

31. Cole, *Element of Risk*, 13–14.

32. Ibid., 14.

33. Ibid.

34. Ibid., 14–16.

35. P. H. Abelson, "Radon Today: The Role of Flimflam in Public Policy," *Regulation* (Fall 1991).

36. Ibid.

37. Ibid.

38. Cole, *Element of Risk*, 15–16.

39. Ibid., 221.

40. UNSCEAR (United Nations Scientific Committee on the Effects of Atomic Radiation), *Sources and Effects of Ionizing Radiation* (New York: United Nations, 1994).

41. Slovic, "Perception of Risk from Radiation," in *Perception of Risk*, ed. Slovic, 264–274.

42. M. deCourcy Hinds, "Radon: Making the Public Pay Attention," *New York Times*, September 24, 1988.

43. W. E. Leary, "13,000 Deaths a Year Indicated by Science Academy Radon Study," *New York Times*, January 6, 1988.

44. J. H. Lubin et al., "Lung Cancer in Radon-Exposed Miners and Estimation of Risk from Indoor Exposure," *Journal of the National Cancer Institute* 87 (June 7, 1995): 817–827; S. Darby, D. Hill, and R. Doll, "Radon: A Likely Carcinogen at All Exposures (review)," *Annals of Oncology* 12 (October 2001): 1341–1351; D. Krewski et al., "Residential Radon and Risk of Lung Cancer: A Combined Analysis of 7 North American Case-Control Studies," *Epidemiology* 16 (March 2005): 137–145.

45. Cole, *Element of Risk*, 170.

46. E. Eckholm, "Radon: The Threat Is Real, but Scientists Argue over Its Severity," *New York Times*, September 2, 1986.

47. Cole, *Element of Risk*, 73–75.

48. Eckholm, "Radon: The Threat Is Real."

49. Ibid.

50. Ibid.

51. Warner et al., "Toward a More Realistic Appraisal."

52. W. W. Nazaroff and K. Teichman, "Indoor Radon: Exploring U.S. Federal Policy for Controlling Human Exposures," *Environmental Science and Technology* 24 (1990): 774–782.

53. Ibid.

54. Abelson, "Radon Today"; Cole, *Element of Risk*, 12–17.

55. Abelson, "Radon Today."

56. National Research Council, *Health Risks of Radon*, 446-454.

57. Ibid., 78.

58. Ibid., 8.

59. Ibid., 8.

60. Ibid., 9–10.

61. Ibid., Table VII-9, 528.

62. Ibid., Table VII-9, 528.

63. T. Reynolds, "Experts Debate Radon's Cancer Risks," *Journal of the National Cancer Institute* 83 (June 19, 1991): 810–812.

64. National Research Council, *Health Effects of Exposure*, 18.

65. Ibid., 19.

66. Ibid., 19.

67. Lubin et al., "Lung Cancer in Radon-Exposed Miners"; J. H. Lubin and J. D. Boice, Jr., "Lung Cancer Risk from Residential Radon: Meta-analysis of Eight Epidemiologic Studies," *Journal of the National Cancer Institute* 89 (January 1997): 49–57.

68. Lubin et al., "Lung Cancer in Radon-Exposed Miners."

69. Lubin and Boice, "Lung Cancer Risk."

70. Ibid.

71. Lubin et al., "Lung Cancer in Radon-Exposed Miners."

72. Ibid.

73. Lubin and Boice, "Lung Cancer Risk."

74. National Research Council, *Health Effects of Exposure*, 19.
75. Lubin et al., "Lung Cancer in Radon-Exposed Miners."
76. Krewski et al., "Residential Radon."
77. Alderson, "A Creeping Suspicion."
78. Krewski et al., "Residential Radon"; Lubin et al., "Lung Cancer in Radon-Exposed Miners."
79. Alderson, "A Creeping Suspicion"; J. H. Lubin, J. M. Samet, and C. Weinberg, "Design Issues in Epidemiologic Studies of Indoor Exposure to Rn and Risk of Lung Cancer," *Health Physics* 59 (December 1990): 807–817.
80. Abelson, "Radon Today"; P. H. Abelson, "Uncertainties About Health Effects of Radon," *Science* 250 (October 19, 1990): 353.
81. Abelson, "Radon Today."
82. M. Upfal, G. Divine, and J. Siemiatycki, "Design Issues in Studies of Radon and Lung Cancer: Implications of the Joint Effect of Smoking and Radon," *Environmental Health Perspectives* 103 (January 1995): 58–63; J. S. Neuberger and T. F. Gesell, "Residential Radon Exposure and Lung Cancer: Risk in Nonsmokers," *Health Physics* 83 (2002): 1–18.
83. Neuberger and Gesell, "Residential Radon Exposure."
84. Lubin et al., "Design Issues in Epidemiologic Studies."
85. Ibid.
86. Ibid.
87. R. E. Thompson et al., "Case-Control Study of Lung Cancer Risk from Residential Radon Exposure in Worcester County, Massachusetts," *Health Physics* 94 (2008): 228–241.
88. Krewski et al., "Residential Radon"; S. Darby et al., "Radon in Homes and Risk of Lung Cancer: Collaborative Analysis of Individual Data from 13 European Case-Control Studies," *BMJ* 330 (January 29, 2005): 223–228.
89. Dale Sandler, interview with author, August 5, 2006.
90. Neuberger and Gesell, "Residential Radon Exposure."
91. Ibid.
92. D. P. Sandler et al., "Indoor Radon and Lung Cancer Risk in Connecticut and Utah," *Journal of Toxicology and Environmental Health*, Part A 69 (2006): 633–654.
93. Ibid.
94. Ibid.
95. Krewski et al., "Residential Radon."
96. Sandler et al., "Indoor Radon."
97. J. S. Neuberger, interview with author, August 16, 2006.
98. Ibid.
99. Sandler, interview with author.
100. U.S. Environmental Protection Agency, *A Citizen's Guide to Radon: The Guide to Protecting Yourself and Your Family from Radon* (Washington, D.C., 2005).
101. Ibid.
102. Stat Bite, "Causes of Lung Cancer in Nonsmokers," *Journal of the National Cancer Institute* 98 (May 17, 2006): 664. The American Cancer Society is credited as the source of the estimates used in the pie chart. But the original source of the number of lung cancer deaths in never smokers due to radon exposure is the National Research Council (*Health Effects of Exposure*). However, ACS selected

the higher of two figures; furthermore, the number is used without any reference to the many attending assumptions and uncertainties due to the extrapolation from the miner data to exposure in homes. In this way, what is a highly uncertain estimate can take on the status of fact.

103. Neuberger and Gesell, "Residential Radon Exposure."

104. P. Boffetta and D. Trichopoulos, "Cancer of the Lung, Larynx, and Pleura," in *Textbook of Cancer Epidemiology*, eds. H.-O. Adami et al. (New York: Oxford University Press, 2002), 248–280.

105. R. C. Brownson et al., "Epidemiology and Prevention of Lung Cancer in Nonsmokers," *Epidemiologic Reviews* 20 (1998): 218–236.

106. J. D. Boice Jr., "Ionizing Radiation," in *Cancer Epidemiology and Prevention* (3rd ed.), eds. D. Schottenfeld and J. F. Fraumeni Jr. (New York: Oxford University Press, 2006), 259–293.

6. THE CONTROVERSY OVER PASSIVE SMOKING: A Casualty of the "Tobacco Wars"

1. R. Kluger, *Ashes to Ashes: America's Hundred-Year Cigarette War, the Public Health, and the Unabashed Triumph of Phillip Morris* (New York: Vintage Books, 1997), 737–739; M. Crichton, "Aliens Cause Global Warming," Michelin Lecture (Pasadena: Caltech, January 17, 2003); P. L. Bernstein, *Against the Gods: The Remarkable Story of Risk* (New York: Wiley, 1996), 211–213.

2. E. L. Wynder and G. C. Kabat, "Environmental Tobacco Smoke and Lung Cancer: A Critical Assessment," in *Indoor Air Quality*, ed. H. Kasuga (Berlin: Springer-Verlag, 1990), 5–15; N. Mantel, "What Is the Epidemiologic Evidence for a Passive Smoking–Lung Cancer Association?" in *Indoor Air Quality*, ed. H. Kasuga (Berlin: Springer-Verlag, 1990), 341–347; J. C. Bailar, "Passive Smoking, Coronary Heart Disease, and Meta-analysis," *New England Journal of Medicine* 340 (March 1999): 958–959.

3. My account of the background to the nonsmokers' rights movement in the 1970s relies heavily on R. Bayer and J. Colgrove ("Science, Politics, and Ideology in the Campaign Against Environmental Tobacco Smoke," *American Journal of Public Health* 92 (June 2002): 949–954) and Richard Kluger (*Ashes to Ashes*).

4. Quoted in Bayer and Colgrove, "Science, Politics, and Ideology."

5. R. Taylor, F. Najafi, and A. Dobson, "Meta-analysis of Studies of Passive Smoking and Lung Cancer: Effects of Study Type and Continent," *International Journal of Epidemiology* 36 (2007): 1048–1059, doi:10.1093/1je/dym158.

6. T. Hirayama, "Nonsmoking Wives of Heavy Smokers have a Higher Risk of Lung Cancer: A Study from Japan," *BMJ* 282 (January 1981): 183–185.

7. L. Garfinkel, "Time Trends in Lung Cancer Mortality Among Nonsmokers and a Note on Passive Smoking," *Journal of the National Cancer Institute* 66 (June 1981): 1061–1066.

8. J. E. Enstrom, "Rising Lung Cancer Mortality Among Nonsmokers," *Journal of the National Cancer Institute* 62 (April 1979): 755–760.

9. O. Auerbach, E. C. Hammond, and L. Garfinkel, "Changes in Bronchial Epithelium in Relation to Cigarette Smoking, 1955-1960 vs. 1970-1977," *New England Journal of Medicine* 300 (February 1979): 381–385.

10. Garfinkel, "Time Trends in Lung Cancer."

11. Ibid.

12. Ibid.

13. L. Garfinkel, "Passive Smoking and Cancer—American Experience," *Preventive Medicine* 13 (November 1984): 691–697.

14. National Research Council, *Environmental Tobacco Smoke: Measuring Exposures and Assessing Health Effects* (Washington, D.C., 1986); U.S Department of Health and Human Services, *The Health Consequences of Involuntary Smoking, A Report of the Surgeon General* (Rockville, Md., 1986); International Agency for Research on Cancer (IARC), *Monographs on the Evaluation of the Carcinogenic Risks of Chemicals to Humans: Tobacco Smoking*, vol. 38 (Lyon, France: IARC, 1986).

15. IARC Monograph, "*Evaluation of the Carcinogenic Risks.*"

16. U.S. Department of Health and Human Services, "*The Health Consequences.*"

17. U.S. Environmental Protection Agency, *Respiratory Health Effects of Passive Smoking: Lung Cancer and Other Disorders* (Washington, D.C.: U.S. Department of Health and Human Services, 1992). My view of the EPA report was shaped both by having served on the committee convened by the EPA to critically review the draft document and by my familiarity with the scientific literature concerning smoking and lung cancer and the occurrence of lung cancer occurring in never smokers. By 1992, I had published three articles on the topic of ETS and lung cancer, and a fourth appeared in 1995: G. C. Kabat and E. L. Wynder, "Lung Cancer in Nonsmokers," *Cancer* 53 (March 1984): 1212–1221; E. L. Wynder and G. C. Kabat, "Environmental Tobacco Smoke"; G. C. Kabat, "Epidemiologic Studies of the Relationship Between Passive Smoking and Lung Cancer," *Toxicology Forum* (February 20, 1990): 187–201; G. C. Kabat, S. D. Stellman, and E. L. Wynder, "Relation Between Environmental Tobacco Smoke and Lung Cancer in Lifetime Nonsmokers," *American Journal of Epidemiology* 142 (July 1995): 141–148.

18. This was clear from the EPA staff's presentation of the background of the report at the hearings in 1990. In addition, in July 1992, after the second meeting of the committee, I telephoned the chairman Dr. Morton Lippman to express my misgivings that the EPA had essentially marshaled the evidence in support of a predetermined conclusion that ETS was a "known human carcinogen." Lippman responded that he would not disagree with my formulation.

19. For example, see pages 1–2, 2–8, 2–9, 5–67, 6–30, 6–31 of the EPA report (*Respiratory Health Effects*).

20. Garfinkel, "Passive Smoking and Cancer"; M. J. Thun et al., "Trends in Smoking and Mortality from Cigarette Use in Cancer Prevention Studies I (1959 through 1965) and II (1982 through 1988)," in *Changes in Cigarette-Related Disease Risks and their Implication for Prevention and Control, Smoking and Tobacco Control Monograph*, vol. 8 (Washington, D. C.: National Cancer Institute, 1996), 318.

21. Another important point concerns the reliance on levels of cotinine (and other tobacco-specific compounds) in body fluids to compare nonsmokers' exposure to ETS with that incurred by active smokers. This approach has been used, in addition to the results of epidemiologic studies, to estimate the risk in nonsmokers. Studies carried out in the 1980s were interpreted as indicating that nonsmokers exposed to ETS had an exposure (uptake of tobacco smoke) equivalent to a smoker of one cigarette per day, and this figure was used to estimate the risk of

nonsmokers. However, it appears that nonsmokers may metabolize cotinine, and perhaps other compounds in tobacco smoke, differently than active smokers. Specifically, nonsmokers may metabolize these products less efficiently than smokers, presumably because smokers have to adapt to coping with much higher levels. For this reason, it appears that comparisons between nonsmokers with exposure to ETS and active smokers may overstate the formers' exposure if they do not take into account the difference in metabolism. The EPA acknowledged this problem but then went on to make the comparison anyway and to accept the claim that the average passive smoker is exposed to the equivalent of 1 cigarette per day.

22. Although I submitted many of these points to the committee (all members were required to hand in written comments on the draft of the document), none made their way into the final version of the report.

23. G. B. Gori and J. C. Luik, *Passive Smoke: The EPA's Betrayal of Science and Policy* (Vancouver: The Fraser Institute, 1999).

24. "The idea that secondhand smoke is a deadly health hazard dovetails so well with the goal of discouraging smoking that tobacco's opponents generally have not been inclined to scrutinize it very closely." J. Sullum, *For Your Own Good: The Anti-Smoking Crusade and the Tyranny of Public Health* (New York: The Free Press, 1998), 160.

25. Bayer and Colgrove, "Science, Politics, and Ideology."

26. P. Brennan et al., "Secondhand Smoke Exposure in Adulthood and Risk of Lung Cancer Among Never Smokers: A Pooled Analysis of Two Large Studies," *International Journal of Cancer* 109 (2004): 125–131.

27. B. Meier, "Judge Voids Study Linking Cancer to Secondhand Smoke," *New York Times*, July 20, 1998. Gori and Luik, *Passive Smoke*, contains a complete transcript of the legal decision.

28. R. C. Brownson et al., "Environmental Tobacco Smoke: Health Effects and Policies to Reduce Exposure," *Annual Review of Public Health* 18 (1997): 163–185.

29. Stat Bite, "Exposure to Secondhand Smoke Among Nonsmokers, 1988–2002," *Journal of the National Cancer Institute* 98 (March 2006): 302.

30. W. C. Hinds and M. W. First, "Concentrations of Nicotine and Tobacco Smoke in Public Places," *New England Journal of Medicine* 292 (April 1975): 844–845.

31. Melvin First, interview with author, November 21, 2003.

32. R. A. Jenkins et al., "Exposure to Environmental Tobacco Smoke in Sixteen Cities in the United States as Determined by Personal Breathing Zone Air Sampling," *Journal of Exposure Analysis and Environmental Epidemiology* 6 (October–December 1996): 473–502; R. A. Jenkins and R. W. Counts, "Personal Exposure to Environmental Tobacco Smoke: Salivary Cotinine, Airborne Nicotine, and Nonsmoker Misclassification," *Journal of Exposure Analysis and Environmental Epidemiology* 9 (July–August 1999): 352–363.

33. M. R. Guerin, R. A. Jenkins, and B. Tomkins, *The Chemistry of Environmental Tobacco Smoke: Composition and Measurement* (2nd ed.), Indoor Air Research Series (Boca Raton, Fla.: CRC Press, 2000).

34. Roger Jenkins, interview with author, September 30 and November 21, 2003.

35. Ibid.

36. K. Phillips et al., "Assessment of Personal Exposures to Environmental Tobacco Smoke in British Nonsmokers," *Environment International* 20 (1994): 693–712.

37. Roger Jenkins, interview with author.

38. Centers for Disease Control, "Achievements in Public Health, 1900–1999: Decline in Deaths from Heart Disease and Stroke—United States, 1900–1999," *MMWR* 48 (1999), www.cdc.gov/mmwr/preview/mmwrhtml/mm4830al.htm.

39. C. S. Fox et al., "Temporal Trends in Coronary Heart Disease Mortality and Sudden Cardiac Death from 1950 to 1999: The Framingham Heart Study," *Circulation* 110 (2004): 522–527.

40. M. R. Law, J. K. Morris, and N. J. Wald, "Environmental Tobacco Smoke Exposure and Ischaemic Heart Disease: An Evaluation of the Evidence," *BMJ* 315 (October 18, 1997): 973–980; M. Thun, J. Henley, and L. Apicella, "Epidemiologic Studies of Fatal and Nonfatal Cardiovascular Disease and ETS Exposure from Spousal Smoking," *Environmental Health Perspectives* 107 (Supplement 6) (December 1999): 841–846; J. He et al., "Passive Smoking and the Risk of Coronary Heart Disease—A Meta-analysis of Epidemiologic Studies," *New England Journal of Medicine* 340 (March 25, 1999): 920–926.

41. K. Steenland, "Passive Smoking and the Risk of Heart Disease," *Journal of the American Medical Association* 267 (January 1992): 94–99.

42. Bailar, "Passive Smoking."

43. M. E. LeVois and M. W. Layard, "Publication Bias in the Environmental Tobacco Smoke/Coronary Heart Disease Epidemiologic Literature," *Regulatory Toxicology and Pharmacology* 21 (February 1995): 184–191.

44. K. Steenland et al., "Environmental Tobacco Smoke and Coronary Heart Disease in the American Society CPS II Cohort," *Circulation* 94 (August 1996): 622–628.

45. M. J. Thun et al., "Trends in Tobacco Smoking and Mortality from Cigarette Use in Cancer Prevention Studies I (1959 through 1965) and II (1982 through 1988)," in *Changes in Cigarette-Related Disease Risks and Their Implication for Prevention and Control*, Smoking and Tobacco Control Monograph 8 (Washington, D. C.: National Cancer Institute, 1996), 311.

46. J. E. Enstrom and G. C. Kabat, "Environmental Tobacco Smoke and Coronary Heart Disease Mortality in the United States—A Meta-analysis and Critique," *Inhalation Toxicology* 18 (March 2006): 199–210.

47. Law et al., "Environmental Tobacco Smoke Exposure."

48. J. E. Enstrom and G. C. Kabat, "Environmental Tobacco Smoke and Tobacco Related Mortality in a Prospective Study of Californians, 1960–98," *BMJ* 326 (May 17, 2003): 1057–1066.

49. Enstrom, "Rising Lung Cancer Mortality."

50. BBC World News. "Row over Passive Smoking Effect," May 16, 2003, http://www.news.bbc.co.uk/2/hi/health/3026933.stm.

51. Ibid.

52. The published correspondence regarding Enstrom and Kabat paper is to be found in *BMJ* 327 (August 30, 2003): 501–505.

53. Richard Smith, "Comment from the Editor," *BMJ* 327 (August 30, 2003): 505.

54. In our 2006 paper, we presented data from five large surveys showing that, in fact, not all nonsmokers back in the 1960s were exposed to ETS.

55. Enstrom and Kabat, "Environmental Tobacco Smoke."

56. Ibid.

57. In contrast to the attention directed at our 2003 *BMJ* paper, this attempt to perform a rigorous and transparent meta-analysis of ETS and heart disease is rarely cited.

58. International Agency for Research on Cancer, *Tobacco Smoke and Involuntary Smoking. IARC Monographs on the Evaluation of Carcinogenic Risks to Humans*, vol. 83 (Lyon, France: IARC, 2003); California Environmental Protection Agency, *Proposed Identification of Environmental Tobacco Smoke as a Toxic Air Contaminant* (draft) (Air Resources Board, California Environmental Protection Agency, June 2005), www.arb.ca.gov/toxics/ets/dreport/dreport.htm; U.S. Department of Health and Human Services, *The Health Consequences of Involuntary Exposure to Tobacco Smoke: A Report of the Surgeon General* (Washington, D.C.: Department of Health and Human Services; 2006), at http://www.surgeongeneral.gov/library/secondhandsmoke.

59. Bailar, "Passive Smoking."

60. IARC, 2004, 1410.

61. As Roger Jenkins pointed out to me, the most glaring instance of the report's failure to put things in perspective is its estimate of the amount of nicotine emitted into the California air by smokers—40 tons per year—but does not mention the amount emitted by the growing of all the solanaceous vegetables in the Central Valley (these include tomatoes, eggplants, and green peppers). That nicotine would be emitted from the leaves of the plants, as they sit out in the hot sun in the cloudless Central Valley. As Jenkins put it, "Everything has a vapor pressure, and nicotine's, especially if it is in the free base form, is fairly high." Similarly, the report gives a figure of 1,907 tons per year of carbon monoxide due to ETS emissions but fails to give any indication of how this compares with other sources. It turns out that all California's cigarette smokers emit, to a first approximation, about the same amount of carbon monoxide as 7,500 cars. This sounds like a lot, until you realize that there are at least 15 million cars in the state of California. Of course, this leaves out other sources of carbon monoxide, such as power plants, forest fires, wood-burning stoves, volcanoes, etc. Thus, to think of cigarettes as a significant contributor to carbon monoxide pollution is ludicrous. The estimates of 40 tons of nicotine and 1,907 tons of carbon monoxide per year are numbers without any context and thus uninformative.

62. L. A. Bero, S. Glantz and M. K. Hong, "The Limits of Competing Interest Disclosure," *Tobacco Control* 14 (April 2005): 118–126; M. J. Thun, "More Misleading Science from the Tobacco Industry," *BMJ* 327 (2003): E237–238E; UPI, 2003; J. A. Francis, A. K. Shea, and J. M. Samet, "Challenging the Epidemiologic Evidence on Passive Smoking: Tactics of Tobacco Industry Expert Witnesses," *Tobacco Control* 15, Supplement IV (2006): iv68–iv76, doi:10.1136/tc.2005.014241; R. L. Barnes, S. K. Hammond, and S. A. Glantz, "The Tobacco Industry's Role in the 16 Cities Study of Secondhand Tobacco Smoke: Do the Data Support the Stated Conclusions? *Environmental Health Perspectives* 114 (December 2006): 1890–1897.

63. For a detailed account of the attacks directed at James Enstrom, see J. E. Enstrom, "Defending Legitimate Epidemiologic Research: Combatting Lysenko Pseudoscience," *Epidemiologic Perspectives and Innovations* 4 (2007): 11.

64. K. J. Rothman, "Conflict of Interest: The New McCarthyism in Science," *Journal of the American Medical Association* 269 (June 1993): 2782–2784.

65. K. J. Rothman, "Conflict of Interest Policies: Protecting Readers or Censoring Authors? In Reply," *Journal of the American Medical Association* 270 (December 8, 1993): 2684.

66. S. Ungar and D. Bray, "Silencing Science: Partisanship and the Career of a Publication Disputing the Danger of Secondhand Smoke," *Public Understanding of Science* 14 (2005): 5–23.

67. M. Siegel, "IN MY VIEW: It's Time to Talk; Tobacco Control Movement Must Open Itself Up for Discussion with Critics," The Rest of the Story: Tobacco News Analysis and Commentary Blog, posted on September 14, 2006, http://www.tobaccoanalysis.blogspot.com/2006/09/in-my-view-its-time-to-talk-tobacco.html (accessed October 7, 2007).

68. Economic Affairs Committee, House of Lords, 2006.

69. S. Sun, J. H. Schiller and A. F. Gazdar, "Lung Cancer in Never Smokers—A Different Disease," *Nature* 7 (October 2007): 778–790.

CONCLUSION

1. Flossman, E. and P. M. Rothwell, "Commentary: Aspirin and Colorectal Cancer: An Epidemiological Success Story," *International Journal of Epidemiology* 36 (2007): 962–965; G. Kune, "Commentary: Aspirin and Cancer Prevention," *International Journal of Epidemiology*. 36 (2007): 957–959.

2. Jordan, V. C., "SERMs: Meeting the Promise of Multifunctional Medicines," *Journal of the National Cancer Institute* 99 (March 7, 2007): 350–356.

3. World Cancer Research Fund. *Food, Nutrition, Physical Activity, and the Prevention of Cancer*. Washington, D. C.: American Institute for Cancer Research, 2007.

4. Doll, R. and R. Peto, *The Causes of Cancer*. New York: Oxford University Press, 1981.

5. Krieger, N., "Epidemiology and the Web of Causation: Has Anyone Seen the Spider?" *Social Science in Medicine* 39 (1994): 887–903; McMichael, A. J., "Prisoners of the Proximate: Loosening the Constraints on Epidemiology in an Age of Change," *American Journal of Epidemiology* 149 (1999): 887–897; Davey Smith, G. and S. Ebrahim, "Epidemiology—Is It Time to Call It a Day?" *International Journal of Epidemiology* 30 (2001): 1–11; March, D. and E. Susser, "The Eco- in Eco-Epidemiology," *International Journal of Epidemiology* 35 (2006): 1379–1383.

6. Houlston, R. S. and J. Peto, "The Future of Association Studies of Common Cancers," *Human Genetics* 112 (2003): 434–435.

7. Schmidt, C. "SNPs Not Living Up to Promise: Experts Suggest New Approach to Disease ID," *Journal of the National Cancer Institute* 99 (February 7, 2007): 188–189; Ioannidis, 2005; Ioannidis, "Common Genetic Variants for Breast Cancer: 32 Largely Refuted Candidates and Larger Prospects," *Journal of the National Cancer Institute* 98 (October 4, 2006): 1350–1353.

GLOSSARY

I have drawn on the following works: Last, J. L., *Dictionary of Epidemiology* (2001) (JL); Gordis, L., *Epidemiology*, 2004 (LG); National Research Council, *Health Effects of Exposure to Radon*, BEIR VI, 1999 (BEIR VI); and Wikipedia (W); initials in parenthesis indicate direct quotes or paraphrases from these sources.

absolute risk: The probability of an event in population under study, as contrasted with the relative risk (JL).

action level: Upper limit on radon concentration in homes, above which the EPA recommends action to reduce the level. The EPA has set an action level of 4 pCi/L of air.

Georgius Agricola (Latinized name of Georg Bauer) (1494–1555): Sixteenth-century German scholar and scientist known as the "father of mineralogy" for his systematic treatise on mining and metallurgy *De re metallica*, published posthumously in 1556. Agricola noted the occurrence of an unusually high death rate from lung disease among underground metal miners in the Erz Mountains spanning Germany and Czechoslovakia (W; BEIR VI).

alpha particle: Two neutrons and two protons bound as a single particle that is emitted from the nucleus of certain radioactive isotopes in the process of decay or disintegration (BEIR VI).

association: A correlation between an exposure or a characteristic and a disease. Association is a necessary condition of a causal relationship, but many phenomena are associated without one of them causing the other. Hence, the dictum "association does not prove causation."

"availability heuristic": Term coined by the psychologists Daniel Kahneman and Amos Tversky to describe the phenomenon that people's judgment about the likelihood, and hence the importance, of an event is influenced by salient events and experiences that make a strong impression but lead to systematic errors. Kahneman won the Nobel Prize for this work in 2002.

becquerel per meter cubed (Bq/m^3): The concentration of radon in the air is measured in units of becquerels per meter cubed or in picocuries per liter (pCi/L) of air. One picocurie per liter is equivalent to 37 Bq/m^3. One becquerel corresponds to 1 disintegration per second.

bias: Deviation of results or inferences from the truth or processes leading to such deviation. Any trend in the collection, analysis, interpretation, or review of data that can lead to conclusions that are systematically different from the truth (JL).

biological marker, or biomarker: A cellular or molecular indicator of exposure, health effects, or susceptibility. Biomarkers can be used to measure internal dose, biologically effective dose, early biological response, altered structure or function, or susceptibility (JL).

BRCA1 and BRCA2: Rare mutations in genes that cause a greatly increased risk of breast and several other cancers in some families. Such mutations account for only a few percent of breast cancers.

case-control study: Type of study in which cases of a particular disease are identified and a comparison group without the disease (controls) are identified. Information on factors thought to play a role in the disease is obtained from both groups and compared. Case-control studies are particularly useful in studying uncommon diseases.

causality: The relating of causes to the effects they produce. Most epidemiology concerns causality, and several different types of causes can be distinguished. It must be emphasized, however, that epidemiologic evidence by itself is insufficient to establish causality, although it can provide powerful circumstantial evidence (JL).

cohort study: Study in which information relevant to the risk of developing disease is collected from members of a defined population or "cohort." The cohort is then followed for a number of years, and cases (or deaths) of the disease of interest are identified. Factors that are associated with the development of the disease can then be evaluated.

concentration: The level of a substance present in a particular environment or in a biological specimen such as blood. For radon, picocuries per liter of air and becquerels per meter cubed are used; for environmental tobacco smoke, common measurements are in micrograms of particles or nicotine per cubic meter of air (a microgram is one millionth of a gram). Concentration is to be distinguished from *exposure*.

confidence interval: A measure of the reliability of a risk estimate. A 95 percent confidence interval means that 95 times out of 100 the estimated risk will fall within the specified interval.

confounding, confounding factor: Refers to the distortion of an observed association between the factor of interest and a disease by a third factor that is associated with both the study factor and the disease.

congeners (PCB congeners): Structurally similar forms of a compound.

cotinine: A major metabolite of nicotine that can be measured in blood, urine, or saliva. Cotinine levels reflect smoking in the previous 2 days.

criteria of judgment: Set of guidelines that are useful in assessing whether an association is likely to be causal. These include the strength of the association, the consistency of the association observed in different studies, temporality (whether the exposure precedes the occurrence of disease), and biological plausibility.

dichloro-diphenyl-trichloroethane (DDT): Synthetic pesticide widely used in agriculture following World War II. In her 1962 book *Silent Spring*, the biologist Rachel Carson drew attention to the effects of indiscriminate use of DDT on wildlife and possibly human health. DDT was banned for most uses in 1972.

DNA adduct: The result of a toxin's binding to the genetic material and causing damage, which, if not repaired, can theoretically lead to cancer.

dose-response relationship: Refers to change in the risk of disease (response) as exposure to the factor of interest (dose) increases. The stronger the dose response seen between a factor and a disease, the more likely it is that the factor causes the disease. The number of cigarettes smoked per day by smokers shows a classic dose-response relationship with their risk of lung cancer.

ecological study: Study in which information pertaining to populations is correlated with the occurrence of disease.

electric fields: Produced by voltage and are measured in volts per meter. Electric fields are easily shield by trees and buildings.

electromagnetic spectrum: The range of all possible electromagnetic radiation from gamma rays, with very high frequencies and energies, to extremely low frequency power frequency fields that have very low energy levels.

ELF-EMF (extremely low frequency electromagnetic fields): Produced by electric power lines are below 3,000 Hz and have wavelengths of more than 5,000 km.

environmental tobacco smoke (ETS): Mixture of smoke from the burning tip of a cigarette and the smoke exhaled by the smoker. Synonymous terms include "passive smoking," "secondhand tobacco smoke," and "involuntary smoking."

etiology: Science or description of the causes of disease.

excess relative risk: Risk due to exposure to a certain level above the background level.

exposure: The condition of having contact with a physical or chemical agent in such a way that the contact can influence the development of disease.

hertz: Unit of frequency expressed in cycles per second. Frequency is proportional to energy level of electromagnetic radiation and inversely proportional to its wavelength.

Human Genome Project: A 13-year project coordinated by the U.S. Department of Energy and the National Institutes of Health in collaboration with scientists in other countries to identify all of the approximately 20,000–25,000 genes in human DNA and determine the sequences of the 3 billion chemical base pairs that make up human DNA. The project was completed in 2003.

incidence: The number of new cases of a disease within a specified period of time.

information bias: Flaw in measuring exposure or outcome data that results in different quality (accuracy) of information between comparison groups (JL). In a case-control study, the information obtained from cases may be affected by the fact of their having received a diagnosis of a serious illness. This may have prompted them to reflect more intensively on their past exposures in an attempt to account for their illness. Healthy controls have do not have the same incentive.

interaction: The phenomenon of two exposures acting jointly to increase risk to a greater extent than either exposure alone; synergism.

ionizing radiation: Electromagnetic radiation with sufficient energies to dislodge electrons from an atom, thereby producing an ion pair. Ionizing radiation includes X-rays and gamma rays, and alpha particles, which can damage DNA through ionization.

Koch's postulates: Formulated by F. G. Jacob Henle and adapted by Robert Koch in 1877 for use in determining whether a particular microorganism was the cause of a specific disease. (1) The agent must be shown to be present in every case of the disease in pure culture; (2) the agent must not be found in cases of other disease; (3) once isolated, the agent must be capable of producing the disease in experimental animals; and (4) the agent must be recovered from the experimental disease produced (JL).

latency, latent period: Period between the onset of exposure and the appearance of clinically detectable disease. In the case of chronic diseases, the latency period can range from several years to decades.

linear-no-threshold (LNT) model: A model of the damage caused by ionizing radiation which presupposes that the response is linear (i.e., directly proportional to the dose) at all dose levels. Thus LNT asserts that there is no threshold of exposure below which the response ceases to be linear (W).

magnetic fields: Produced by current flowing in wires. Magnetic fields are not easily shielded. Measured in gauss (G) or tesla (T).

meta-analysis: A technique used to combine the results of a number of small studies in order to obtain a summary estimate, which, it is hoped, will better describe the true relationship. Meta-analysis can be carried out using the data available in published papers, in contrast to pooled analyses, which involve reanalyzing the original data from different studies using a common approach.

microtesla (μT): Measure of magnetic fields (specifically flux density). One microtesla is equivalent 10 milligauss.

milligauss (mG): Measure of magnetic fields (flux density). One milligauss = 0.1 microteslas.

misclassification: The erroneous classification of an individual, a value, or an attribute into a category other than that to which it should be assigned.

The probability of misclassification may be the same in all study groups (nondifferential misclassification) or may vary between groups (differential misclassification) (JL).

odds ratio: The measure of association obtained from a case-control study, compares the "odds" of exposure to the factor of interest among the cases to the odds of exposure among the controls.

organochlorine compounds: Class of compounds including DDT, heptachlor, chlordane, mirex, and PCBs (polychlorinated biphenyls).

PAH (polycyclic aromatic hydrocarbon): Class of compounds produced by combustion of fossil fuels and other organic matter.

Paracelsus (adopted name of Philippus Theophrastus Aureolus Bombastus von Hohenheim) (1493–1541): Sixteenth-century Swiss alchemist and physician who wrote, "All things are poison and nothing is without poison, only the dose permits something not to be poisonous." Paracelsus pioneered the use of chemicals and minerals in medicine and engaged in experimentation to learn about the human body. He believed that sickness and health were dependent on the balance of specific minerals and that certain illnesses had chemical remedies that could cure them. Like his contemporary, Agricola, he wrote a major work on mining, *On the Miners' Sickness and Other Diseases of Miners*, in which he documented the occupational hazards of metalworking as well as treatment and prevention strategies. (W)

picocurie: One trillionth of a curie. A curie is the amount of radioactivity emitted by the decay of a gram of radium. Picocuries per liter of air (pCi/L) is a measure of radon concentration or level in a given environment.

polychlorinated biphenyls (PCBs): Class of organic compounds that are no longer produced the United States but are still found in the environment. PCBs were widely used in the electrical and other industries, but were banned in the 1970s due to their toxicity.

pooled analysis: The collaborative reanalysis of data from a number of studies using a common analytic approach.

prevalence: Frequency of occurrence of a factor in the population.

prospective study: Another term for a cohort study.

radiation: Energy emitted in the form of waves or particles by radioactive atoms as a result of radioactive decay (BEIR VI).

radionuclide: An atom with an unstable nucleus, which undergoes radioactive decay, emiting ionizing radiation. (BEIR VI).

radon: A naturally occurring radioactive gas produced by the decay of radium, which in turn is produced by the decay of uranium; decays to form radon progeny (or radon "daughters") (BEIR VI).

radon progeny or daughters: The radioactive products formed following the radioactive decay of radon; radionuclides which when inhaled can expose living cells to their emitted alpha particles (BEIR VI).

Ramazzini, Bernadino (1633–1714): Seventeenth-century Italian physician and founder of the field of occupational medicine. He was an early proponent of the use of cinchona bark (from which quinine is derived) in the treatment of malaria. His most important contribution to medicine was his work on occupational diseases called *De Morbis Artificum Diatriba* (*Diseases of Workers*) which outlined the health hazards of chemicals, dusts, metals, and other agents encountered by workers in fifty-two occupations (W). In this work, he noted that breast cancer occurred more often in nuns then in the general population of women and speculated that some aspect of childbearing might be protective.

relative risk: The ratio of the risk of disease or death among the exposed to the risk among the unexposed (JL).

risk: The probability that an event will occur, e.g., the probability that an individual will become ill or die within a stated period of time or by certain age. Also a nontechnical term encompassing a variety of measures of the probability of a (generally) unfavorable outcome (JL).

risk factor: A personal characteristic or exposure that in an epidemiologic study is associated with the occurrence of disease. The characteristic could be a sociodemographic factor (such as age, sex, ethnicity), a clinical characteristic (cholesterol level), an exposure, or a genetic factor. Often the term is used to imply a causal relationship, when this is not appropriate. Age is a strong risk factor for colon cancer and many other diseases, but age is not a cause of colon cancer. It is more appropriate to interpret "risk factor" to mean "risk marker."

selection bias: Error due to systematic differences in characteristics between those who take part in a study and those who do not. Selection bias can invalidate conclusions and generalizations that might otherwise be drawn from such studies (JL).

selective estrogen receptor modulators (SERMs): A class of drugs that act on the estrogen receptor to selectively inhibit or stimulate estrogen-like action in various tissues. Tamoxifen and raloxifene are examples of SERMs.

statistical significance: A measure of whether a particular result is unlikely to be due to chance. If the 95 percent confidence interval associated with a relative risk does not include 1.0, the result is judged to be statistically significant.

tesla: Measure of magnetic fields, specifically of magnetic flux density. One tesla is equivalent to 10,000 gauss (G). Magnetic fields in homes are measured in microTesla, or μT.

threshold: Level below which an agent or toxin has no adverse effect.

working level (WL): Measure of the concentration of radon daughters, which is a measure of the potential alpha particles energy per liter of air. One WL of radon daughters corresponds approximately to a level of 200 pCi/L of radon in a typical indoor environment.

working level months (WLM): A measure of cumulative exposure to radon used in studies of miners and equivalent to 1 working level for a working month (170 hr) (BEIR VI).

BIBLIOGRAPHY

Abelson, P. H. "Radon Today: The Role of Flimflam in Public Policy." *Regulation* (Fall 1991).

——. "Uncertainties About Health Effects of Radon." *Science* 250 (1990): 353.

Adair, E. R. "'Currents of Death' Rectified." New York: IEEE-USA, 1991.

Adair, R. K. "Biological Responses to Weak 60-Hz Electric and Magnetic Fields Must Vary as the Square of the Field Strength." *Proceedings of the National Academy of Sciences, USA* 91 (1994): 9422–9425.

——. "Constraints on Biological Effects of Weak Extremely-Low-Frequency Electromagnetic Fields." *Physical Review A* 43 (1991): 1039–1048.

——. "A Critique of the NIEHS Working Group Report on Assessment of Health Effects from Exposure to Power-Line Frequency Electric and Magnetic Fields" (September 17, 1998).

——. "Fear of Weak Electromagnetic Fields," *Scientific Review of Alternative Medicine* 3 (1999): 22–23; http://www.skeptically.org/quackery/id1.html (accessed March 30, 2008).

——. "Measurements Described in a Paper by Blackman, Blanchard, Bename, and House Are Statistically Invalid." *Bioelectromagnetics* 17 (1996): 510–511.

——. "Static and Low-Frequency Magnetic Fields: Health Risks and Therapies." *Reports on Progress in Physics* 63 (2000): 415–454.

Adami, H.-O., L. Lipworth, L. Titus-Ernstoff, C.-C. Hsieh, A. Hanberg, U. Ahlborg, J. Baron, and D. Trichopoulos. "Organochlorine Compounds and Estrogen-Related Cancers in Women." *Cancer Causes and Control* 6 (1995): 551–566.

Ahlbom, A., N. Day, M. Feychting, R. E. Skinner, J. Dockerty, M. Linet, M. McBride, J. Michaelis, J. H. Olsen, T. Tynes, and P. K. Verkasalo. "A Pooled Analysis of Magnetic Fields and Childhood Leukemia." *British Journal of Cancer* 83 (2000): 692–698.

Ahlborg, U. G., L. Lipworth, L. Titus-Ernstoff, C.-C. Hsieh, A. Hanberg, J. Baron, and D. Trichopoulos. "Organochlorine Compounds in Relation to Breast Cancer, Endometrial Cancer, and Endometriosis: An Assessment of the Biological and Epidemiological Evidence." *Critical Reviews in Toxicology* 25 (1995): 463–531.

Alderson, L. "A Creeping Suspicion About Radon." *Environmental Health Perspectives* 102 (October 1994): 826–831.

Ames, B. N. and L. S. Gold. "Paracelsus to Parascience: The Environment and Cancer Distraction." *Mutation Research* 447 (2000): 3–13.

Ames, B. N., M. Profet, and L. S. Gold. "Dietary Pesticides (99.99% All Natural)." *Proceedings of the National Academy of Sciences, USA* 87 (1990): 7777–7781.

Auerbach, O., E. C. Hammond, and L. Garfinkel. "Changes in Bronchial Epithelium in Relation to Cigarette Smoking, 1955–1960 vs. 1970–1977." *New England Journal of Medicine* 300 (February 1979): 381–385.

Bailar, J. C. "Passive Smoking, Coronary Heart Disease, and Meta-analysis." *New England Journal of Medicine* 340 (March 1999): 958–959.

Bain, C. W., W. Willett, B. Rosner, F. E. Speizer, C. Belanger, and C. H. Hennekens. "Early Age at First Birth and Decreased Risk of Breast Cancer." *American Journal of Epidemiology* 114 (1981): 705–709.

Barnes, R. L., S. K. Hammond, and S. A. Glantz. "The Tobacco Industry's Role in the 16 Cities Study of Secondhand Tobacco Smoke: Do the Data Support the Stated Conclusions?" *Environmental Health Perspectives* 114 (December 2006): 1890–1897.

Bayer, R. and J. Colgrove. "Science, Politics, and Ideology in the Campaign Against Environmental Tobacco Smoke." *American Journal of Public Health* 92 (June 2002): 949–954.

BBC World News, "Row over Passive Smoking Effect." May 16, 2003. http://news.bbc.co.uk/2/hi/health/3026933.stm (accessed October 7, 2007).

Beasley, R. P. "Hepatitis B Virus: The Major Etiology of Hepatocellular Carcinoma." *Cancer* 61 (1988): 1942–1956.

Bennett, W. R., Jr. "Cancer and Power Lines." *Physics Today* (April 1994): 23–29.

Bernstein, P. L. *Against the Gods: The Remarkable Story of Risk*. New York: Wiley, 1996.

Bero, L. A., S. Glantz, and M. K. Hong. "The Limits of Competing Interest Disclosure." *Tobacco Control* 14 (April 2005): 118–126.

Boffetta, P., and D. Trichopoulos. "Cancer of the Lung, Larynx, and Pleura," in *Textbook of Cancer Epidemiology*, edited by H.-O. Adami, D. Hunter, and D. Trichopoulos, 248–280. New York: Oxford University Press, 2002.

Boffetta, P., J. K. McLaughlin, C. La Vecchia, P. Autier, and P. Boyle. "'Environment' in Cancer Causation and Etiological Fraction: Limitations and Ambiguities." *Carcinogenesis* 28 (May 2007): 913–915.

Boice, J. D., Jr., "Ionizing Radiation," in *Cancer Epidemiology and Prevention*, edited by D. Schottenfeld and J. F. Fraumeni Jr., 259–293. New York: Oxford University Press, 2006.

Boice, J. D., Jr., and J. K. McLaughlin. "Epidemiologic Studies of Cellular Telephone and Cancer Risk: A Review." *SSI Report (Statens strålskyddsinstitut)* 16 (2002).

Bracken, M. B., K. Belanger, K. Hellenbrand, K. Addesso, S. Patel, E. Triche, and B. P Leaderer. "Correlates of Residential Wiring Code Used in Studies of Health Effects of Residential Electromagnetic Fields." *American Journal of Epidemiology* 148 (1998): 467–474.

Brainard, G. C., R. Kavet, and L. I. Kheifets. "The Relationship Between Electromagnetic Field and Light Exposures to Melatonin and Breast Cancer Risk: A Review of the Relevant Literature." *Journal of Pineal Research* 26 (1999): 65–100.

Brennan, P., P. A. Buffler, P. Reynolds, A. H. Wu, E. Wichmann, A. Agudo, G. Pershagen, K-H. Jöckel, S. Benhamou, R. S. Greenberg, F. Merletti, C. Winck, E. T. H. Fontham, M. Kreuzer, S. C. Darby, F. Forastière, L. Simonata, and P. Boffetta. "Secondhand Smoke Exposure in Adulthood and Risk of Lung Cancer Among Never Smokers: A Pooled Analysis of Two Large Studies." *International Journal of Cancer* 109 (2004): 125–131.

Brodeur, P. *Currents of Death: Power Lines, Computer Terminals, and the Attempt to Cover Up Their Threat to Your Health.* New York: Simon and Shuster, 1989.

Bromley, D. A. *The President's Scientist: Reminiscences of a White House Science Advisor.* New Haven: Yale University Press, 1994.

Brownson, R. C., M. P. Eriksen, R. M. Davis, and K. E. Warner. "Environmental Tobacco Smoke: Health Effects and Policies to Reduce Exposure." *Annual Review of Public Health* 18 (1997): 163–185.

Brownson, R. C., M. C. R. Alavanja, N. Caporaso, E. J. Simoes, and J. C. Chang. "Epidemiology and Prevention of Lung Cancer in Nonsmokers." *Epidemiologic Reviews*, 20 (1998): 218–236.

Butler, T. L., "The Relationship of Passive Smoking to Various Health Outcomes Among Seventh Day Adventists in California." Doctoral dissertation, University of California, Los Angeles, 1988.

California Environmental Protection Agency. *Proposed Identification of Environmental Tobacco Smoke as a Toxic Air Contaminant.* Air Resources Board, June 2005, www.arb.ca.gov/toxics/ets/dreport/dreport.htm.

Calle, E. E., H. Frumkin, J. Henley, D. A. Savitz, and M. J. Thun. "Organochlorines and Breast Cancer Risk." *CA A Cancer Journal for Clinicians* 52 (2002): 301–309.

Campion, E. W. "Power Lines, Cancer, and Fear." *New England Journal of Medicine* 337 (1997): 44–46.

"Cancer Risk from Domestic Radon." *Lancet* 1 (January 14, 1989): 93.

Carroll, K. K., and H. T. Khor. "Dietary Fat in Relation to Tumorigenesis." *Progress in Biochemical Pharmacology* 10 (1975): 308–353.

Centers for Disease Control, "Achievements in Public Health, 1900–1999: Decline in Deaths from Heart Disease and Stroke—United States, 1900-1999." *MMWR.* http://www.cdc.gov/mmwr/preview/mmwrhtml/mm4830al.htm.

Cho, E., S. A. Smith-Warner, J. Ritz, P. van den Brandt, G. A. Colditz, A. R. Folsom, J. L. Freudenheim, E. Giovannucci, A. Goldbohm, S. Graham, L. Holmberg, D. H. Kim, N. Malila, A. B. Miller, P. Pietinen, T. E. Rohan,

T. A. Sellers, F. E. Spiezer, W. C. Willett, A. Wolk, and D. J. Hunter. "Alcohol Intake and Colorectal Cancer: A Pooled Analysis of 8 Cohort Studies." *Annals of Internal Medicine* 140 (2004): 603–613.

Colborn, T., D. Dumanoski, and J. P. Myers. *Our Stolen Future: Are We Threatening our Fertility, Intelligence, and Survival?* New York: Plume, 1996.

Colditz, G. A. "Relationship Between Estrogen Levels, Use of Hormone Replacement Therapy, and Breast Cancer." *Journal of the National Cancer Institute* 90 (1998): 814-823.

Colditz, G. A., W. C. Willett, D. J. Hunter, M. J. Stampfer, J. E. Manson, C. H. Hennekens, and B. A. Rosner. "Family History, Age, and Risk of Breast Cancer: Prospective Data from the Nurses' Health Study." *Journal of the American Medical Association* 270 (1993): 338–343.

Cole, L. A. *Element of Risk: The Politics of Radon.* New York: Oxford University Press, 1993.

Cole, P., D. Trichopoulos, H. Pastides, T. Starr, and J. S. Mandel. "Dioxin and Cancer: A Critical Review." *Regulatory Toxicology and Pharmacology* 38 (2003): 378–388.

Crichton, M. "Aliens Cause Global Warming." Michelin Lecture. Pasadena: Caltech, January 17, 2003.

Cummings, K. M., A. Brown, and R. O'Connor. "The Cigarette Controversy." *Cancer Epidemiology Biomarkers and Prevention* 16 (June 2007): 1070–1076.

Darby, S., D. Hill, and R. Doll. "Radon: A Likely Carcinogen at All Exposures." *Annals of Oncology* 12 (October 2001): 1341–1351.

Darby, S., D. Hill, J. M. Barros-Dios, H. Baysson, F. Bochicchio, H. Deo, R. Falk et al. "Radon in Homes and Risk of Lung Cancer: Collaborative Analysis of Individual Data from 13 European Case-Control Studies." *BMJ* 330 (January 29, 2005): 223–228.

Davey Smith, G., and S. Ebrahim. Epidemiology—Is It Time to Call It a Day?" *International Journal of Epidemiology* 30 (2001): 1–11.

Davis, S., D. K. Mirick, and R. G. Stevens. "Residential Magnetic Fields and the Risk of Breast Cancer." *American Journal of Epidemiology* 155 (2002): 446–454.

deCourcy Hinds, M. "Radon: Making the Public Pay Attention." *New York Times*, September 24, 1988.

Derby, C. A., H. A. Lamont, M McFarland Barbour, J. B. McPhillips, T. M. Lasater, and R. A Carleton. "Correlates of Postmenopausal Estrogen Use and Trends Through the 1980s in Two Southeastern New England Communities." *American Journal of Epidemiology* 137 (1993): 1125–1135.

Doll, R., and R. Peto. *The Causes of Cancer.* New York: Oxford University Press, 1981.

Douglas, M., and A. Wildavsky. *Risk and Culture: An Essay on the Selection of Technological and Environmental Dangers.* Berkeley: University of California Press, 1982.

Eckholm, E. "Radon: The Threat Is Real, but Scientists Argue over Its Severity." *New York Times*, September 2, 1986.

Economic Affairs Committee, House of Lords, "Memorandum Submitted by Professor Andrew Evans," http://www.publications.parliament.uk/pa/ld200506/ldselect/ldeconaf/999/econ140206.pdf (accessed April 25, 2006).

Egger, M., M. Schneider, and G. Davey Smith. "Spurious Precision? Meta-analysis of Observational Studies." *BMJ* 316 (1998): 140–144.

Elwood, J. M., J. C. G. Pearson, D. H. Skipper, and S. M. Jackson. "Alcohol, Smoking, and Social and Occupational Factors in the Aetiology of Cancer of the Oral Cavity, Pharynx, and Larynx." *International Journal of Cancer* 34 (1984): 603–612.

Enstrom, J. E. "Defending Legitimate Epidemiologic Research: Combating Lysenko Pseudoscience." *Epidemiologic Perspectives and Innovations* 4 (October 10, 2007): 11.

——. "Rising Lung Cancer Mortality Among Nonsmokers." *Journal of the National Cancer* Institute 62 (April 1979): 755–760.

Enstrom, J. E., and G. C. Kabat. "The Authors Respond." *BMJ* 327 (August 30, 2003): 501–505.

——. "Environmental Tobacco Smoke and Coronary Heart Disease Mortality in the United States—A Meta-analysis and Critique." *Inhalation Toxicology* 18 (March 2006): 199–210.

——. "Environmental Tobacco Smoke and Tobacco Related Mortality in a Prospective Study of Californians, 1960–98." *BMJ* 326 (May 17, 2003): 1057–1066.

Epstein, S. S. "A Straw Man," *New York Times*, October 22, 1988.

Ernster, V. L., L. Mason, W. H. Goodson 3rd, E. A. Sickles, S. T. Sacks, S. Selvin, M. E. Dupuy, J. Hawkinson, and T. K. Hunt. "Effects of Caffeine-Free Diet on Benign Breast Disease: A Randomized Trial." *Surgery* 91 (1982): 263–267.

Fagin, D. "So Many Things Went Wrong: Costly Search for Links Between Pollution and Breast Cancer Was Hobbled from the Start." *Newsday*, July 29, 2002.

——. "Tattered Hopes: A $30-Million Federal Study of Breast Cancer and Pollution on LI Has Disappointed Activists and Scientists." *Newsday*, July 28, 2002.

Ferraro, S. "You Can't Look Away Anymore: The Anguished Politics of Breast Cancer." *New York Times Magazine*, August 15, 1993.

Feychting, M., and A. Ahlbom. "Childhood Leukemia and Residential Exposure to Weak Extremely Low Frequency Magnetic Fields." *Environmental Health Perspectives*, suppl. 2 (1995): 59–62.

——. "Magnetic Fields and Cancer in Children Residing near Swedish High-Voltage Power Lines." *American Journal of Epidemiology* 138 (1993): 467–481.

Feyerabend, P. *Against Method*, London: Verso, 2002.

Flegal, K. M., B. I Graubard, D. F. Williamson, and M. H. Gail. "Cause-Specific Excess Deaths Associated with Underweight, Overweight, and Obesity." *Journal of the American Medical Association* 298 (November 7, 2007): 2028–2037.

Flossman, E., and P. M. Rothwell. "Commentary: Aspirin and Colorectal Cancer: An Epidemiological Success Story." *International Journal of Epidemiology*. 36 (2007): 962–965.

Foucault, M. *Power/Knowledge: Selected Interviews and Other Writings, 1972–77*, ed. Colin Gordon. New York: Pantheon, 1980.

Fox, C. S., J. C. Evans, M. G. Larson, W. B. Kannel, and D. Levy. "Temporal Trends in Coronary Heart Disease Mortality and Sudden Cardiac Death from 1950 to 1999: The Framingham Heart Study." *Circulation* 110 (2004): 522–527.

Francis, J. A., A. K. Shea, and J. M. Samet. "Challenging the Epidemiologic Evidence on Passive Smoking: Tactics of Tobacco Industry Expert Witness." *Tobacco Control* 15, suppl. IV (2006):iv68–iv76, doi:10.1136/tc.2005.014241.

Gammon, M. D., M. S. Wolff, A. I. Neugut, S. M. Eng, S. L. Teitelbaum, J. A. Britton, M. B. Terry, et al. "Environmental Toxins and Breast Cancer on Long Island. II. Organochlorine Compound Levels in Blood." *Cancer Epidemiology Biomarkers and Prevention* 11 (2002): 686–697.

Gammon, M. D., and R. M. Santella. "Reply to F. Perrera." *Cancer Epidemiology Biomarkers and Prevention* 12 (2003): 75–76.

Garfinkel, L. "Passive Smoking and Cancer—American Experience." *Preventive Medicine* 13 (November 1984): 691–697.

——. "Time Trends in Lung Cancer Mortality Among Nonsmokers and a Note on Passive Smoking." *Journal of the National Cancer Institute* 66 (June 1981): 1061–1066.

Garland, C.;, E. Barrett-Conner, L. Suarez, M. H. Criqui, and D. L. Wingard. "Effects of Passive Smoking on Ischemic Heart Disease Mortality on Nonsmokers: a Prospective Study," *American Journal of Epidemiology* 121 (1985): 645–650.

Gawande, A. "The Cancer Cluster Myth," *The New Yorker*, February 8, 1999.

Gilliland, F. D., W. C. Hunt, V. E. Archer, and G. Saccomanno. "Radon Progeny Exposure and Lung Cancer Risk Among Non-Smoking Uranium Miners." *Health Physics* 79 (October 2000): 365–372.

Goldsmith, M. F. "How Serious Is the Indoor Radon Health Hazard?" *Journal of the American Medical Association* 258 (August 7, 1987): 578–579.

Gordis, L. *Epidemiology*, 3rd ed. Philadelphia: Elsevier Saunders, 2004.

Gori, G. B., and J. C. Luik. *Passive Smoke: The EPA's Betrayal of Science and Policy*. Vancouver: The Fraser Institute, 1999.

Grady, D., "Pregnancy Problems Tied to Caffeine." *New York Times*, January 21, 2008.

Grady, D., T. Gebretsadik, K. Kerlikowske, V. Ernster, and D. Petitti. "Hormone Replacement Therapy and Endometrial Cancer Risk: A Meta-analysis." *Obstetrics and Gynecology* 85 (1995): 304–313.

Graham, J. D., K. Clemente, and R. Glass. "Breast Cancer: What Are the Perceived Risk Factors?" *Harvard Center for Risk Analysis* 4 (1996): 1–2.

Greenland, S., A. R. Sheppard, W. T. Kaune, C. Poole, and M. A. Kelsh. "A Pooled Analysis of Magnetic Fields, Wire Codes, and Childhood Leukemia." *Epidemiology* 11 (2000): 624–634.

Grodstein, F., P. A. Newcomb, and M. J. Stampfer. "Postmenopausal Hormone Therapy and the Risk of Colorectal Cancer: A Review and Meta-Analysis." *American Journal of Medicine* 106 (1999): 574–582.

Guerin, M. R., R. A. Jenkins, and B. Tomkins. *The Chemistry of Environmental Tobacco Smoke: Composition and Measurement*, 2nd ed., Indoor Air Research Series. Boca Raton, Fla.: CRC, 2000.

Hackshaw, A. K., M. R. Law, and N. J. Wald. "The Accumulated Evidence on Lung Cancer and Environmental Tobacco Smoke." *BMJ* 315 (1997): 980–988.

Hardell, L., M. J. Walker, B. Walhjalt, L. S. Friedman, and E. D. Richter. "Secret Ties to Industry and Conflicting Interests in Cancer Research." *American Journal of Industrial Medicine* (2007) 50: 227–233.

Harris, G. "For F.D.A., A Major Backlog Overseas." *New York Times*, January 29, 2008.

Harris, G., and A. O'Connor. "On Autism's Cause, It's Parents vs. Research." *New York Times*, June 25, 2005.

Hatch, E., R. A. Kleinerman, M. S. Linet, R. E. Tarone, W. T. Kaune, A. Auvinen, D. Baris, L. Robison, and S. Wacholder. "Do Confounding or Selection Factors of Residential Wiring Codes and Magnetic Fields Distort Findings of Electromagnetic Fields Studies." *Epidemiology* 11 (2000): 189–198.

He, J., S. Vupputuri, K. Allen, M. R. Prerost, J. Hughes, and P. K. Whelton. "Passive Smoking and the Risk of Coronary Heart Disease—A Meta-analysis of Epidemiologic Studies." *New England Journal of Medicine* 340 (March 25, 1999): 920–926.

Hill, A. B. "The Environment and Disease: Association or Causation?" *Proceedings of the Royal Society of Medicine* 58 (1965): 295–300.

Hinds, W. C., and M. W. First. "Concentrations of Nicotine and Tobacco Smoke in Public Places." *New England Journal of Medicine* 292 (April 1975): 844–845.

Hirayama, T. "Nonsmoking Wives of Heavy Smokers Have a Higher Risk of Lung Cancer: a Study from Japan." *British Medical Journal* 282 (January 1981): 183–185.

Holford, T. R., K. A. Cronin, A. B. Mariotto, and E. J. Feuer. "Changing Patterns in Breast Cancer Incidence Trends." *Journal of the National Cancer Institute Monographs*. no. 36 (2007): 19–25.

Houlston, R. S., and J. Peto. "The Future of Association Studies of Common Cancers." *Human Genetics* 112 (2003): 434–435.

Humble, C., J. Croft, A. Gerber, M. Casper, C. G. Hames, and H. A. Tyroler. "Passive Smoking and 20-Year Cardiovascular Disease Mortality Among Nonsmoking Wives, Evans County, Georgia," *American Journal of Public Health* 80 (1990): 599–601.

Hunter, D. J., and K. T. Kelsey. "Pesticide Residues and Breast Cancer: The Harvest of a Silent Spring?" *Journal of the National Cancer Institute* 85 (April 21, 1993): 598–599.

Institute of Medicine. Committee to Review the Health Effects in Vietnam Veterans of Exposure to Herbicides. *Veterans and Agent Orange: Health Effects of Herbicides Used in Vietnam.* Washington, D.C.: National Academy of Sciences Press, 1994.

International Agency for Research on Cancer. *IARC Monograph 44: Alcohol Drinking*. Lyon: IARC, 1988.

——. *Tobacco Smoke and Involuntary Smoking*. Monographs on the Evaluation of Carcinogenic Risks to Humans, vol. 83. Lyon: IARC, 2003.

——. *Tobacco Smoking*, Monographs on the Evaluation of the Carcinogenic Risks of Chemical to Humans, vol. 38. Lyon: IARC, 1986.

Ioannidis, J. P. A. "Common Genetic Variants for Breast Cancer: 32 Largely Refuted Candidate and Larger Prospects." *Journal of the National Cancer Institute* 98 (October 4, 2006): 1350–1353.

——. "Molecular Bias." *European Journal of Epidemiology* 20 (2005):739–745.

——. "Why Most Published Research Findings Are False." *PLoS Medicine* 2 (August 2005): e124.

Jenkins, R. A., and R. W. Counts. "Personal Exposure to Environmental Tobacco Smoke: Salivary Cotinine, Airborne Nicotine, and Nonsmoker Misclassification." *Journal of Exposure Analysis and Environmental Epidemiology* 9 (July–August 1999): 352–363.

Jenkins, R. A., A. Palausky, R. W. Counts, C. K. Bayne, A. B. Dindal, and M. R. Guerin. "Exposure to Environmental Tobacco Smoke in Sixteen Cities in the United States as Determined by Personal Breathing Zone Air Sampling." *Journal of Exposure Analysis and Environmental Epidemiology* 6 (October–December 1996): 473–502.

Jenks, S. "Researchers to Comb Long Island for Potential Cancer Factors." *Journal of the National Cancer Institute* 86 (1994): 88–89.

Johannes, C. B., S. L. Crawford, J. G. Posner, and S. M. McKinlay. "Longitudinal Patterns and Correlates of Hormone Replacement Therapy Use in Middle-Aged Women." *American Journal of Epidemiology* 140 (1994): 439–452.

Jordan, V. C. "SERMs: Meeting the Promise of Multifunctional Medicines." *Journal of the National Cancer Institute* 99 (March 7, 2007): 350–356.

Kabat, G. C. "Epidemiologic Studies of the Relationship Between Passive Smoking and Lung Cancer." *Toxicology Forum* (February 20, 1990): 187–201.

Kabat, G. C., and E. L. Wynder. "Lung Cancer in Nonsmokers." *Cancer* 53 (March 1984): 1212–1221.

Kabat, G. C., S. D. Stellman, and E. L. Wynder. "Relation Between Environmental Tobacco Smoke and Lung Cancer in Lifetime Nonsmokers." *American Journal of Epidemiology* 142 (July 1995): 141–148.

Kabat, G. C., E. S. O'Leary, E. R. Schoenfeld, J. M. Greene, R. Grimson, K. Henderson, W. T. Kaune et al. "Electric Blanket Use and Breast Cancer on Long Island." *Epidemiology* 14 (2003): 514–520.

Kahneman, D., and A. Tversky. "Judgment Under Uncertainty: Heuristics and Biases." *Science* 185 (September 1974): 1124–1131.

Kawachi, I., G. A. Colditz, F. E. Speizer, J. E. Manson, M. J. Stampfer, W. C. Willett, and C. H. Hennekens. "A Prospective Study of Passive Smoking and Coronary Heart Disease," *Circulation* 95 (1997): 2374–2379.

Key, T., and G. Reeves. "Organochlorines in the Environment and Breast Cancer." *BMJ* 308 (June 11, 1994): 1520–1521.

Kheifets, L. I., R. S. Greenberg, R. R. Neutra, G. L. Hester, C. L. Poole, D. P. Rall, and G. Lundell. "Electric and Magnetic Fields and Cancer: Case Study." *American Journal of Epidemiology* 154 (2000): S50–S59.

Kluger, R. *Ashes to Ashes: America's Hundred-Year Cigarette War, the Public Health, and the Unabashed Triumph of Philip Morris*. New York: Random House, 1997.

Kolata, G. "Environment and Cancer: The Links Are Elusive." *New York Times*, December 13, 2005.

——."The Epidemic That Wasn't." *New York Times*, August 29, 2002.

——."For Radiation, How Much Is Too Much?" *New York Times*, November 27, 2001.

——."Hormones and Cancer: Assessing the Risks." *New York Times*, December 26, 2006.

——."Reversing Trend, Big Drop Is Seen in Breast Cancer." *New York Times*, December 15, 2006.

Krewski. D., J. H. Lubin, J. M. Zielinksi, M. Alavanja, V. S. Catalan, R. W. Field, J. B. Klotz et al. "Residential Radon and Risk of Lung Cancer: A Combined Analysis of 7 North American Case-Control Studies." *Epidemiology* 16 (March 2005): 137–145.

Krieger, N. "Epidemiology and the Web of Causation: Has Anyone Seen the Spider?" *Social Science in Medicine* 39 (1994): 887–903.

Kuhn, T. S. *The Structure of Scientific Revolutions*, 2nd ed. Chicago: University of Chicago Press, 1970.

Kulldorff, M., E. J. Feuer, B. A. Miller, and L. S. Freedman. "Breast Cancer Clusters in the Northeast United States: A Geographic Analysis." *American Journal of Epidemiology* 146 (July 15, 1997): 161–170.

Kuller, L. "Is Phenomenology the Best Approach to Health Research?" *American Journal of Epidemiology* 166 (2007): 1109–1115.

Kune, G. "Commentary: Aspirin and Cancer Prevention." *International Journal of Epidemiology* 36 (2007): 957–959.

Laden, F., S. E. Hankinson, M. S. Wolff, G. A. Colditz, W. C. Willett, F. E. Speizer, and D. J. Hunter. "Plasma Organochlorine Levels and the Risk of Breast Cancer: An Extended Follow-Up in the Nurses' Health Study." *International Journal of Cancer* 91 (2001): 568–574.

Last, J. M. *A Dictionary of Epidemiology*, 4th ed. New York: Oxford University Press, 2001.

Law, M. R., J. K. Morris, and N. J. Wald. "Environmental Tobacco Smoke Exposure and Ischaemic Heart Disease: An Evaluation of the Evidence." *BMJ* 315 (October 18, 1997): 973–980.

Leary, W. E. "13,000 Deaths a Year Indicated by Science Academy Radon Study." *New York Times*, January 6, 1988.

LeVois, M. E., and M. W. Layard. "Publication Bias in the Environmental Tobacco Smoke/Coronary Heart Disease Epidemiologic Literature." *Regulatory Toxicology and Pharmacology* 21 (February 1995): 184–191.

Levy, D. T., E. A. Mumford, K. M. Cummings, E. A. Gilpin, G. Giovino, A. Hyland, D. Sweanor, and K. E. Warner. "The Relative Risks of a Low-Nitrosamine Smokeless Tobacco Product Compared with Smoking Cigarettes: Estimates of a Panel of Experts," *Cancer Epidemiology Biomarkers and Prevention* 13 (2004): 2035–2042.

Linet, M. S., E. E. Hatch, R. A. Kleinerman, L. L. Robison, W. T. Kaune, D. R. Friedman, R. K. Severson et al. "Residential Exposure to Magnetic Fields and Acute Lymphoblastic Leukemia in Children." *New England Journal of Medicine* 337 (1997): 1–7.

London, S. J., D. C. Thomas, J. D. Bowman, E. Sobel, T. C. Cheng, and J. M. Peters. "Exposure to Residential Electric and Magnetic Fields and Risk of Childhood Leukemia." *American Journal of Epidemiology* 134 (1991): 923–937.

London, S. J., J. M. Pogoda, K. L. Hwang, B. Langholz, K. R. Monroe, L. N. Kolonel, W. T. Kaune, J. M. Peters, and B. E. Henderson. "Residential Magnetic Field Exposure and Breast Cancer Risk: A Nested Case-Control Study from a Multiethnic Cohort in Los Angeles County, California." *American Journal of Epidemiology* 158 (2003): 969–980.

Lopez-Cervantes, M., L. Torres-Sanchez, A. Tobias, and L. Lopez-Carillo. "Dichlorodiphenyldichloroethane Burden and Breast Cancer Risk: A Meta-analysis of the Epidemiologic Evidence." *Environmental Health Perspectives* 112 (2004): 207–214.

Lubin, J. H. and J. D. Boice Jr. "Lung Cancer Risk from Residential Radon: Meta-analysis of Eight Epidemiologic Studies." *Journal of the National Cancer Institute* 89 (January 1, 1997): 49–57.

Lubin, J. H., J. M. Samet, and C. Weinberg. "Design Issues in Epidemiologic Studies of Indoor Exposure to Rn and Risk of Lung Cancer." *Health Physics* 59 (December 1990): 807–817.

Lubin, J. H., J. D. Boice Jr., C. Edling, R. W. Hornung, G. R. Howe, E. Kunz, R. A. Kusiak et al. "Lung Cancer in Radon-Exposed Miners and Estimation of Risk from Indoor Exposure." *Journal of the National Cancer Institute* 87 (June 7, 1995): 817–827.

Lucas, R. M., and A. J. McMichael. "Association or Causation: Evaluating Links Between "Environment and Disease." *Bulletin of the World Health Organization* 83 (2005): 792–795.

MacMahon, B., S. Yen, D. Trichopoulos, K. Warren, and G. Nardi. "Coffee and Cancer of the Pancreas." *New England Journal of Medicine* 304 (March 1981): 630–633.

Mantel, N. "What Is the Epidemiologic Evidence for a Passive Smoking-Lung Cancer Association?" In *Indoor Air Quality*, ed. H. Kasuga. Berlin: Springer-Verlag, 1990: 341–347.

March, D., and E. Susser. "The Eco- in Eco-Epidemiology." *International Journal of Epidemiology* 35 (2006): 1379–1383.

Markowitz, G., and D. Rosner. *Deceit and Denial: The Deadly Politics of Industrial Pollution*. Berkeley: University of California Press, 2002.

Marks, P. "U.S. to Finance Project to Study Breast Cancer on Long Island," *New York Times*, November 25, 1993.

Marmot, M. *The Status Syndrome: How Social Standing Affects Our Health and Longevity*. London: Bloomsbury, 2005.

Matthews, K. A, L. H. Kuller, R. R. Wing, E. N. Meilahn, and P. Plantinga. "Prior Use of Estrogen Replacement Therapy, Are Users Healthier Than Nonsusers?" *American Journal of Epidemiology* 143 (May 15, 1996): 971–978

McMichael, A. J. "Prisoners of the Proximate: Loosening the Constraints on Epidemiology in an Age of Change." *American Journal of Epidemiology* 149 (1999): 887–897.

McNeil, C. "Breast Cancer Decline Mirrors Fall in Hormone Use, Spurs Both Debate and Research." *Journal of the National Cancer Institute*, 99(4) (February 21, 2007): 266–267.

Meier, B. "Judge Voids Study Linking Cancer to Secondhand Smoke," *New York Times*, July 20, 1998.

Merton, R. K. *The Sociology of Science: Theoretical and Empirical Investigations*, Chicago: University of Chicago Press, 1973.

Michaels, D., and C. Monforton. "Manufacturing Uncertainty: Contested Science and the Protection of the Public's Health and Environment." *American Journal of Public Health* 95 (2005): S39–S48.

Mooney, C. *The Republican War on Science*. New York: Basic Books, 2005.

Morabia, A. "On the Origins of Hill's Causal Criteria." *Epidemiology* 2 (1991): 367–369.

——. "Risky Concepts: Methods in Cancer Research." *American Journal of Public Health* 91 (2001): 355–357.

Morris, J. N., "Uses of Epidemiology." *International Journal of Epidemiology* 36 (2007): 1165–1172.

Muñoz, N., X. Castellsagué, A. Berrington de Gonzalez, and L. Gissmann. "HPV in the Etiology of Human Cancer." *Vaccine* S3 (2006): S1–S10.

National Institute of Environmental Health Services *NIEHS Report on Health Effects from Exposure to Power-Line Frequency Electric and Magnetic Fields.* NIH Publication no. 99-4493, Washington D. C., 1999.

National Research Council, *Environmental Tobacco Smoke: Measuring Exposures and Assessing Health Effects.* Washington, D. C.: National Academy Press, 1986.

——. *Health Effects of Exposure to Radon*, BEIR VI. Washington, D. C.: National Academy Press, 1999.

——. *Health Risks of Radon and Other Internally-Deposited Alpha-Emitters.* BEIR IV. Washington, D. C.: National Academy Press, 1988.

——. *Possible Health Effects of Exposure to Residential Electric and Magnetic Fields.* Washington, D. C.: National Academy Press, 1997.

Nazaroff, W. W., and K. Teichman. "Indoor Radon: Exploring U.S. Federal Policy for Controlling Human Exposures." *Environmental Science and Technology* 24 (1990): 774–782.

Neuberger, J. S., and T. F. Gesell. "Residential Radon Exposure and Lung Cancer: Risk in Nonsmokers." *Health Physics* 83 (2002): 1–18.

Newton, M., and A. L. Young. "The Story of 2,4,5-T: A Case Study of Science and Societal Concerns." *Environmental Science and Pollution Research* 11 (2004): 207–208.

Niederdeppe, J., and A. G. Levy. "Fatalistic Beliefs About Cancer Prevention and Three Prevention Behaviors." *Cancer Epidemiology Biomarkers and Prevention* 16 (May 1, 2007): 988–1003, doi:10.1158/1055-9965.EPI-06-0608.

Ochs, R. "The LI Breast Cancer Study." *Newsday*, December 10, 1996.

ORAU Panel on Health Effects of Low-Frequency Electric and Magnetic Fields. "EMF and Cancer." *Science* 260 (April 2, 1993): 14–16.

Park, R. L. *Voodoo Science: The Road from Foolishness to Fraud*. New York: Oxford University Press, 2000.

Pear, R. "Safety Standard is Set on Radon in U.S. Homes," *New York Times*, August 15, 1986.

Pershagen, G., G. Akerblom, O. Axelson, B. Clavensjo, L. Damber, G. Desai, A. Enflo, F. Lagarde, H. Mellander, M. Svartengren, and G. A. Swedjemark. "Residential Radon Exposure and Lung Cancer in Sweden." *New England Journal of Medicine* 330 (1994): 159–164.

Peto, J. "Cancer Epidemiology in the Last Century and the Next Decade." *Nature* 411 (2001): 390–395.

Phillips, C. V. "Warning: Anti-Tobacco Activism May Be Hazardous to Epidemiologic Science." *Epidemiologic Perspectives and Innovations* 4 (2007): 13, doi:10.1186/1742-5573-4-13.

Phillips, C. V., and K. J. Goodman. "The Missed Lessons of Sir Austin Bradford Hill." *Epidemiologic Perspectives & Innovations* 1 (October 2004): 3, doi:10.1186/1742-5573-1-3.

Phillips, D. H. "DNA Adducts As Markers of Exposure and Risk." *Mutation Research* 577 (2005): 284–292.

Phillips, K., D. A. Howard, D. Browne, and J. M. Lewsley. "Assessment of Personal Exposures to Environmental Tobacco Smoke in British Nonsmokers." *Environment International* 20 (1994): 693–712.

Pidgeon, N., R. E. Kasperson, and P. Slovic (eds). *The Social Amplification of Risk.* Cambridge: Cambridge University Press, 2003.

Poole, C., and D. Trichopoulos. "Extremely Low-Frequency Electric and Magnetic Fields and Cancer." *Cancer Causes and Control* 2 (1991): 267–276.

Pope, C. A., III, M. J. Thun, M. M. Namboodiri, D. W. Dockery, J. S. Evans, F. E. Speizer, and C. W. Health Jr. "Particulate Air Pollution As a Predictor of Mortality in a Prospective Study of U.S. Adults." *American Journal of Respiratory and Critical Care Medicine* 151 (1995): 669–674.

Proctor, R. N. *Cancer Wars: How Politics Shapes What We Know and Don't Know about Cancer.* New York: Basic Books, 1995.

Ravdin, P. M., K. A. Cronin, N. Howlander, C. D. Berg, R. T. Chlebowski, E. J. Feuer, B. K. Edwards, and D. A. Berry. "The Decrease in Breast-Cancer Incidence in 2003 in the United States." *New England Journal of Medicine.* 356 (April 19, 2007): 1670–1674.

Rest, K. M., and M. H. Halpern. "Politics and the Erosion of Federal Scientific Capacity: Restoring Scientific Integrity to Public Health Science." *American Journal of Public Health* 97 (2007): 1939–1944.

Revkin A. C. "Bush vs. the Laureates: How Science Became a Partisan Issue." *New York Times*, October 19, 2004.

Reynolds, T. "Experts Debate Radon's Cancer Risks." *Journal of the National Cancer Institute* 83 (June 19, 1991): 810–812.

Ropeik, D., and G. Gray. *Risk! A Practical Guide for Deciding What's Really Safe and What's Really Dangerous in the World Around You.* Boston: Houghton Mifflin, 2002.

Rothman, K. J. "Conflict of Interest: The New McCarthyism in Science." *Journal of the American Medical Association* 269 (June 1993): 2782–2784.

——. "Conflict of Interest Policies: Protecting Readers or Censoring Authors? In Reply." *Journal of the American Medical Association* 270 (December 8, 1993): 2684.

——. *Epidemiology: An Introduction.* New York: Oxford University Press, 2002.

Rothman, K. J., and S. Greenland. "Causation and Causal Inference in Epidemiology." *American Journal of Public Health* 95 (2005): S144–S150.

Rothman, K., and A. Keller. "The Effect of Joint Exposure to Alcohol and Tobacco on Risk of Cancer of the Mouth and Pharynx." *Journal of Chronic Disease* 25 (December 1972): 711–716.

Safe, S. H. "Is There an Association Between Exposure to Environmental Estrogens and Breast Cancer?" *Environmental Health Perspectives*, 105, suppl. 3 (April 1997): 675–678.

Sandler, D. P. "On Blankets and Breast Cancer." *Epidemiology* 14 (2003): 509.

Sandler, D. P., K. J. Helsing, G. W. Comstock, and D. L. Shore. "Deaths from All Causes in Non-smokers who Lived with Smokers," *American Journal of Public Health* 129 (1989): 380–387.

Sandler, D. P., C. R. Weinberg., D. L. Shore, V. E. Archer, M. B. Stone, J. L. Lyon, L. Rothney-Kozlak, M. Shepherd, and J. A. J. Stolwijk. "Indoor Radon and Lung Cancer Risk in Connecticut and Utah." *Journal of Toxicology and Environmental Health* Part A, 69 (2006): 633–654.

Sanghavi, D. "Treat Me: The Crucial Health Stat You've Never Heard of." Slate, posted September 26, 2006, www.slate.com/toolbar.aspx?action=print&id=2150354.

Savitz, D. A. "Health Effects of Electric and Magnetic Fields: Are We Done Yet?" *Epidemiology* 14 (2003): 15–17.

Savitz, D. A., and A. Ahlbom. "Electromagnetic Fields and Radiofrequency Radiation." In *Cancer Epidemiology*, eds. D. Schottenfeld and J. R. Frameni Jr. Oxford: Oxford University Press, 2006: 306–321.

Savitz, D. A., and E. E. Calle. "Leukemia and Occupational Exposure to Electromagnetic Fields: Review of Epidemiologic Surveys." *Journal of Occupational Medicine* 29 (1987): 47–51.

Savitz, D. A., R. L. Chan, A. H. Herring, P. P. Howards, and K. E. Hartmann. "Case-Control Study of Childhood Cancer and Exposure to 60-Hz Magnetic Fields." *American Journal of Epidemiology* 128 (1988): 21–38.

Savitz, D. A., N. E. Pearce, and C. Poole. "Methodological Issues in the Epidemiology of Electromagnetic Fields and Cancer." *Epidemiologic Reviews* 11 (1989): 59–78.

Schoenfeld, E. R., E. S. O'Leary, K. Henderson, R. Grimson, G. C. Kabat, S. Ahn, W. T. Kaune, M. D. Gammon, and M. C. Leske. "Electromagnetic Fields and Breast Cancer on Long Island: A Case-Control Study." *American Journal of Epidemiology* 158 (2003): 47–58.

Schmidt, C. "SNPs Not Living Up to Promise; Experts Suggest New Approach to Disease ID." *Journal of the National Cancer Institute* 99 (February 7, 2007): 188–189.

Schwartz, L. M., S. Woloshin, E. L. Dvorin, and H. G. Welsh. "Ratio Measures in Leading Medical Journals: Structured Review of Accessibility of Underlying Absolute Risks." *BMJ* 333 (October 2006): 1248, doi:10.1136/bmj,38985.564317.7C.

Shabecoff, P. "Drive to Locate Risk Areas for Radioactive Gas Urged." *New York Times*, May 24, 1985.

——. "E.P.A. Proposes 5-Year Program Aimed at Radioactive Radon Gas." *New York Times*, October 10, 1985.

——. "Issue of Radon: New Focus on Ecology." *New York Times*, September 10, 1986.

——. "Radioactive Gas in Soil Raises Concern in Three-State Area." *New York Times*, May 19, 1985.

Shah, N. R., J. Borenstein, and R. W. Dubois. "Postmenopausal Hormone Therapy and Breast Cancer: A Systematic Review and Meta-analysis." *Menopause* 12 (2005): 653–655.

Shane, S. "Debating the Evidence of Gulf War Illnesses." *New York Times*, November 16, 2004.

Shapiro, S. "Meta-analysis/Shmeta-analysis." *American Journal of Epidemiology* 140 (1994): 771–778.

Siegel, M. "The Rest of the Story: Tobacco News Analysis and Commentary," tobaccoanalysis.blogspot.com/2006/09/in-my-view-its-time-to-talk-tobacco.html (accessed October 7, 2007).

Slovic, P., ed. *The Perception of Risk*. London: Earthscan, 2000.

——. "Perception of Risk from Radiation." In *The Perception of Risk*, 264–274. London: Earthscan, 2000.

Smith, R. "Comment from the Editor," *BMJ* 327 (August 30, 2003): 505.

Sokoloff, H. "Cancer Map of LI Unveiled." *Newsday*, September 21, 2000.

Stat Bite. "Causes of Lung Cancer in Nonsmokers." *Journal of the National Cancer Institute* 98 (May 17, 2006): 664.

——. "Exposure to Secondhand Smoke Among Nonsmokers, 1988–2002." *Journal of the National Cancer Institute* 98 (March 1, 2006): 302.

Steenland, K. "Passive Smoking and the Risk of Heart Disease." *Journal of the American Medical Association* 267 (January 1992): 94–99.

Steenland, K., M. Thun, C. Lally, and C. Heath. "Environmental Tobacco Smoke and Coronary Heart Disease in the American Society CPS II Cohort." *Circulation* 94 (August 1996): 622–628.

Stein, Z. A. "Silicone Breast Implants: Epidemiological Evidence of Sequelae." *American Journal of Public Health* 89 (1999): 484–487.

Stevens, R. G. "Electric Power Use and Breast Cancer: A Hypothesis." *American Journal of Epidemiology* 125 (1987): 556–561.

Sturgeon, S. R., C. Schairer, M. Gail, M. McAdams, L. A. Brinton, and R. N. Hoover. "Geographic Variation in Mortality from Breast Cancer Among White Women in the United States." *Journal of the National Cancer Institute* 87 (December 20, 1995): 1846–1853.

Sullum, J. *For Your Own Good: The Anti-Smoking Crusade and the Tyranny of Public Health*. New York: The Free Press, 1998.

Sun, S., J. H. Schiller, and A. F. Gazdar. "Lung Cancer in Never Smokers—A Different Disease." *Nature* 7 (October 2007): 778–790.

Sunstein, C. R. *Risk and Reason: Safety, Law, and the Environment*. Cambrdige: Cambridge University Press, 2002.

Susser, M. *Causal Thinking in the Health Sciences: Concepts and Strategies of Epidemiology*. New York: Oxford University Press, 1973.

Svendsen, K. H., L. H. Kuller, M. J. Matin, and J. K. Ockene. "Effect of Passive Smoking in the Multiple Risk Factor Intervention Trial," *American Journal of Epidemiology* 126 (1987): 783–795.

Taubes, G. "Epidemiology Faces Its Limits." *Science* 269 (July 14, 1995): 164–169.

——. "Unhealthy Science: Why Can't We Trust Much of What We Hear About Diet, Health, and Behavior-Related Diseases?" *New York Times Magazine*, September 16, 2007.

Taylor R., F. Najafi, and A. Dobson. "Meta-analysis of Studies of Passive Smoking and Lung Cancer: Effects of Study Type and Continent." *International Journal of Epidemiology* 36 (2007): 1048–1059, doi:10.1093/1je/dym158.

Thompson, R. E., D. F. Nelson, J. H. Popkin, and Z. Popkin. "Case-Control Study of Lung Cancer Risk from Residential Radon Exposure in Worcester County, Massachusetts." *Health Physics* 94 (2008): 228–241.

Thun, M. J., C. Day-Lally, D. G. Myers, E. E. Calle, D. W. Flanders, B.-P. Zhu, M. M. Namboodiri, and C. W. Heath. "Trends in Tobacco Smoking and Mortality from Cigarette Use in Cancer Prevention Studies I (1959 Through 1965) and II (1982 Through 1988)." In *Changes in Cigarette-Related Disease Risks and Their Implication for Prevention and Control*, Smoking and Tobacco Control Monograph 8. Washington, D. C.: National Cancer Institute, 1996: 305–382.

Thun, M., J. Henley, and L. Apicella. "Epidemiologic Studies of Fatal and Nonfatal Cardiovascular Disease and ETS Exposure from Spousal Smoking." *Environmental Health Perspectives* 107, suppl. 6 (December 1999): 841–846.

Thun, M. J. "More Misleading Science from the Tobacco Industry." *BMJ* 327 (2003): E237–238E.

Tomenius, L. "50-Hz Electromagnetic Environment and the Incidence of Childhood Tumors in Stockholm County." *Biolectromagnetics* 7 (1986): 191–207.

Ungar, S., and D. Bray. "Silencing Science: Partisanship and the Career of a Publication Disputing the Danger of Secondhand Smoke." *Public Understanding of Science* 14 (2005): 5–23.

United Nations Scientific Committee on the Effects of Atomic Radiation. *Sources and Effects of Ionizing Radiation*. New York: United Nations, 1994.

Upfal, M., G. Divine, and J. Siemiatycki. "Design Issues in Studies of Radon and Lung Cancer: Implications of the Joint Effect of Smoking and Radon." *Environmental Health Perspectives* 103 (January 1995): 58–63.

U.S. Department of Health and Human Services. "Findings of Scientific Misconduct." *NIH Guide*, June 18, 1999.

——. *The Health Consequences of Involuntary Exposure to Tobacco Smoke: A Report of the Surgeon General.* Rockville, Md., 2006. http:// www.surgeongeneral.gov/library/secondhandsmoke.

——. *The Health Consequences of Involuntary Smoking, A report of the Surgeon General.* Rockville, Md., 1986.

U.S. Environmental Protection Agency, *A Citizen's Guide to Radon: The Guide to Protecting Yourself and Your Family from Radon*, Washington, D.C., 2005.

——. *A Citizen's Guide to Radon: What It Is and What to Do About It.* Washington, D. C., August 1986.

——. *Respiratory Health Effects of Passive Smoking: Lung Cancer and Other Disorders.* Rockville, Md.: U.S. Department of Health and Human Services, 1992.

——. *Technical support document for the 1992 Citizen's Guide to Radon*, Report EPA 400-R-92-011. Washington, D.C., 1992.

U.S. Public Health Service. *Smoking and Health: Report of the Advisory Committee to the Surgeon General of the Public Health Service*, Publication no. 1103. Washington, D.C.: U.S. Department of Health, Education, and Welfare, Public Health Service, Center for Disease Control, 1964.

"U.S. Says Radon Gas Is States' Fight." *New York Times.* August 16, 1986.

Wald, M. L. "With New Data, A Debate on Low-Level Radiation." *New York Times*, July 7, 2005.

Warner, K. E., D. D. Mendez, and P. N. Courant. "Toward a More Realistic Appraisal of the Lung Cancer Risk from Radon: The Effects of Residential Mobility." *American Journal of Public Health* 86 (September 1, 1996): 1222–1227.

Weaver, J. C., and R. D. Astumian. "Response of Cells to Very Weak Electric Fields." *Science* 247 (1990): 459–462.

Wertheimer, N., and E. Leeper. "Electrical Wiring Configurations and Childhood Cancer." *American Journal of Epidemiology* 109 (1979): 273–284.

Willett, W. C. "Prospective Studies of Diet and Breast Cancer." *Cancer* 74, suppl. 3 (1994): 1085–1089.

Wilson, R., and A. Shlyakhter. "Re: Magnetic Fields and Cancer in Children Residing near Swedish High-Voltage Power Lines." *American Journal of Epidemiology* 141 (1995): 378–379.

Wolff, M. S., P. G. Toniolo, E. W. Lee, M. Rivera, and N. Dubin. "Blood Levels of Organochlorine Residues and Risk of Breast Cancer." *Journal of the National Cancer Institute* 85 (1993): 648–652.

Wolff, M. S., A. Zeleniuch-Jacquotte, N. Dubin, and P. Toniolo. "Risk of Breast Cancer and Organochlorine Exposure." *Cancer Epidemiology Biomarkers Prevention* 9 (2000): 271–277.

World Cancer Research Fund. *Food, Nutrition, Physical Activity, and the Prevention of Cancer.* Washington, D. C.: American Institute for Cancer Research, 2007.

Woznicki, Katrina. "Smoking Effects Still Debated." *United Press International*, June 23, 2003.

Wright, Karen. "Going by the Numbers." *New York Times Magazine*, December 15, 1991.

Wynder, E. L., and G. C. Kabat. "Environmental Tobacco Smoke and Lung Cancer: A Critical Assessment." In *Indoor Air Quality*, ed. H. Kasuga. Berlin: Springer-Verlag, 1990: 5–15.

Wynder, E. L., I. T. T. Higgins, and R. E. Harris. "The Wish Bias." *Journal of Clinical Epidemiology* 43 (1990): 619–621.

CREDITS

Grateful acknowledgement is made for permission to reprint or adapt the following:

FIGURES

Figure 2.1 By permission of S. Karger A.G., Basel.

Figure 3.1 Copyright © 2007 Massachusetts Medical Society. All rights reserved.

Figure 3.2 By permission of Oxford University Press.

Figure 3.3 Reproduced with permission from Environmental Health Perspectives.

Figure 4.1 By permission of Oxford University Press.

Figure 5.1 By permission of Oxford University Press.

Figure 5.2 Reprinted with permission from the National Academies Press. Copyright © 1999 National Academy of Sciences.

Figure 5.3 Reproduced with permission from the BMJ Publishing Group.

Figure 5.6 By permission of Oxford University Press.

Figure 6.1 Courtesy of R. A. Jenkins.

TABLES

Table 2.4 Reprinted with permission from Elsevier.

Table 4.1 By permission of Oxford University Press.

Table 4.2 By permission of Oxford University Press.

Table 4.3 Copyright © 1997 Massachusetts Medical Society. All rights reserved.

Table 5.1 By permission of Lippincott Williams and Wilkins.

Table 6.1 Reprinted by permission of Taylor & Francis Ltd.

Table 6.2 Reprinted by permission of Taylor & Francis Ltd.

INDEX